THE NOON FIX

JAMES N. WILSON

AuthorHouse™
1663 Liberty Drive, Suite 200
Bloomington, IN 47403
www.authorhouse.com
Phone: 1-800-839-8640

First published by AuthorHouse 3/30/2009
Revised version published by AuthorHouse 6/1/2009

ISBN: 978-1-4389-5866-8 (sc)

Printed in the United States of America
Bloomington, Indiana

This book is printed on acid-free paper.

INTRODUCTION

When I first tried the double altitude method to determine longitude, I quickly found that north-south boat movement affected the time of maximum altitude. I immediately started looking for a simple, direct method of compensating for it. When none was forthcoming, I wrestled with the mathematics to derive the equation for the time between maximum altitude and meridian transit (Reference 1). That was a bit messy, but it worked. But recently, the simple, direct method hit me. Like many others, I slapped my head, muttering, "Why didn't I think of this before?"

The approach is intended as a backup for GPS, and is for navigators who are familiar with the process of sight taking and reduction. Accordingly, only differences from that are described in detail.

THE METHOD

The double altitude approach averages the times for equal altitudes before and after meridian passage. But north-south observer movement and declination change need to be considered, since they affect the measured altitudes. From reference 1, (Sn - d) is the rate of movement between the observer and the body, where Sn is the northerly component of speed, and d is the rate of change of declination, here positive if the change is northerly. Multiplying this by the time between observations gives the resultant change in altitude.

At meridian transit the navigational triangle has become a line. At this time, the change in sextant altitude (Δhs) is (Sn - d)ΔWT, where ΔWT is the difference in watch times between observations. It can be simply added to or subtracted from the initially measured altitude hs. Averaging the times of the initial hs and the adjusted second hs gives the time of meridian transit. That's it!

Relying on single observations is not recommended, but it does illustrate the basic approach. Reference 1 describes a graphical method for determining the time of maximum altitude. On this plot, adjust the last altitude line by Δhs. Figure 2 illustrates this, where line [3] is adjusted to line [4], whose intersection with line [1] is at the time of meridian transit.

AN EXAMPLE

A modification of the graphical approach described in reference 1 will be used here. A meridian transit sight reduction form has been created to assist the user. Working between the form and the plot will ease the burden. The data are the same as in reference 1.

Start with the form, Figure 1. Enter the basic data, including the estimated dead reckoning (DR) position at meridian transit. From that the estimated time of transit can be calculated in the usual manner, after which declination can be calculated in the declination box. Then switch to the plot, Figure 2.

After the first run of sights, plot hs vs WT. Start plot at 11-28-00; fair a straight line through the points (Figure 2, line 1).

Enter the WT of the last sight at the bottom of the Time of Transit box, and subtract it from the estimated time of transit above. Then add the difference to that time. This is the approximate time to start the last series of sights.

After the second run, plot hs at the top of the page, using a compressed WT scale. Fair a curve through these points (Figure 2, curve 2) to determine the maximum hs for latitude determination.

While waiting for the altitude to decrease, calculate the latitude. Make the corrections in the altitude box, and determine the latitude in the latitude box, as usual. The diagram to the left of that box is for convenience, to help visualize the relationships.

After the third run of sights, establish a second WT scale, beginning at 12-18-00. Plot this run, using different symbols to avoid confusion with the first run, again fairing a straight line through the points (Figure 2, line 3).

Use the LAN (Local Apparent Noon) box to first determine Sn, using the graph on the form. Subtract d, noting that in the declination box, declination is S and decreasing. Thus declination is changing northerly, and d is positive. On the plot, near the estimated time of transit, calculate ΔWT as (81 - 31) = 50 minutes. From the Increments and Corrections pages in the Nautical Almanac, (Yellow or grey tinted pages) for (Sn - d) of -5.4'/hr, determine Δhs as -4.5'.

Measure off -4.5', and draw line 4 parallel to line 3.

Average 11-30 and 12-20 to get 11-55, and read the time below the intersection of lines 1 and 4 as 11-55-43. This is the time of meridian transit. Convert this to UT, and determine longitude in the Longitude box.

PREPARATION

Aside from knowing where all the tools are to use this method, at least one watch aboard should be treated like a chronometer, i. e., a log should be kept of its error and its rate. A practice fix determination under good conditions will make life easier for the navigator should an emergency arise.

DISCUSSION

The intent of this method was to simplify the approach. But three unexpected bonuses resulted. The first is that the method is insensitive to the body selected. The same approach can be used to get the time of meridian transit for all bodies, including the moon. The second is that the asymmetry of the altitude-time curve due to north-south motion is automatically considered. The third is that changes in longitude are also automatically considered. These benefits accrue because the method utilizes the data contained in the slope of the altitude lines. Consider an observer with a westward component of movement. His measured altitudes will rise and fall at a lesser rate than an observer traveling along a meridian. Thus the hs versus WT lines will have a flatter slope. This means that there will be a greater WT difference between equal altitudes. Thus the resultant change in hs will be greater, and the time difference between maximum altitude and meridian transit will also be greater.

ERROR SENSITIVITY

The primary source of error is in the sights themselves, or the plotting of them and the fairing of the lines through the plotted points.

ΔWT is dependent on time of meridian transit and vice versa. Using the estimated time of transit based on DR longitude introduces a very small error, unless the final time differs significantly from that.

To be exact, the hs for latitude determination should be taken at LAN instead of at maximum altitude, but the resultant error is quite small.

Sn is properly speed over the ground. The effect of current and leeway should be considered. If course and speed vary during the interval, using the average is satisfactory.

If the index correction (IC) has changed from the first run to the third, the difference can be applied to Δhs.

A more detailed error discussion is contained in reference 1.

CONCLUSION

While the method is aimed at getting the simplest backup to GPS, other uses may materialize. That it has the capability for more accuracy than the equation in reference 1 (and its predecessor in the *Admiralty Navigation Manual* some fifty plus years earlier) could be of benefit.

ACKNOWLEDGMENTS

Invaluable help was provided by Elliot R. (Joe) Cutting of JPL, who continually offered encouragement and critical analyses; by Frederick H. (Ham) Wright of the Pasadena Power Squadron, for pressing for the use of the graphical approach; and by Paul M. Janiczek of Navigation, for urging the use of graphics instead of equations. Without their contributions, this method might never have materialized.

REFERENCE

1. *Position from Observation of a Single Body*, James N. Wilson, *Navigation*, Volume 32, Number 1, Spring

MERIDIAN TRANSIT SIGHT REDUCTION FORM

TIME OF TRANSIT

DR GHA _____ 118 ° 16 . 6 '
= Lo W, 360° - Lo E

GHA _19_ h _104_ ° _21_ . _0_ '

GHA Diff _____ 13 ° 55 . 6 '

Min-Sec _____ 55 - 42

UT _____ 19 - 55 - 42

ZD $^{E-}_{W+}$ (+) _____ 8 _____ (rev)

ZT _____ 11 - 55 - 42

- - - - - - - - - - - - - - - -

Last sight WT _____ 11 - 33 - 12

Difference _____ 22 - 30

ZT + Difference _____ 12 - 18 - 12

LAN

S __6__ . __0__ kt   Cn __210__ °

Sn (−) _____ 5 . 2 kt

Sn - d (−) _____ 5 . 4 '/hr

d positive if change is Northerly

ΔWT __50__ m

Δ hs (−) _____ 4 . 5 '

WT _____ 11 - 55 - 43

WE $^{f-}_{s+}$ ( ) _____ 0

ZT _____ 11 - 55 - 43

ZD $^{E-}_{W+}$ (+) __8__

UT _____ 19 - 55 - 43

LONGITUDE

GHA _19_ h __104__ ° __21__ . __0__ '

_55_ m _43_ s __13__ ° __55__ . __8__ '

Lo __118__° __16__ . __8__ ' Ⓦ

360° __359__° __59__ . __10__ '

Lo _____ ° _____ . _____ ' E

BASIC DATA

Date _30 Dec 1982_ Body _Sun LL_

DR L _____ 33 ° 40 . 0 ' Ⓝ S E W

DR Lo _____ 118 ° 16 . 6 ' N S E Ⓦ

DECLINATION

Dec _19_ hr __23__ ° __09__ . __1__ ' N Ⓢ

d (−) __0__ . __2__ '/hr

d corr (−) _____ 0 . 2 '

Dec _____ 23 ° 08 . 9 ' N Ⓢ

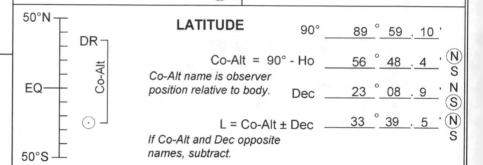

ALTITUDE

Ht of eye _____ 6 . 0 ft

hs _____ 32 ° 57 . 7 '

IC (+) _____ 1 . 5 '

Dip (−) _____ 2 . 4 '

Total (−) _____ 0 . 9 '

ha _____ 32 ° 56 . 8 '

Corr (+) _____ 14 . 8 '

Ho _____ 33 ° 11 . 6 '

LATITUDE

90° _____ 89 ° 59 . 10 '

Co-Alt = 90° - Ho _____ 56 ° 48 . 4 ' Ⓝ S

Co-Alt name is observer position relative to body. Dec _____ 23 ° 08 . 9 ' N Ⓢ

L = Co-Alt ± Dec _____ 33 ° 39 . 5 ' Ⓝ S

If Co-Alt and Dec opposite names, subtract.

Sn DETERMINATION

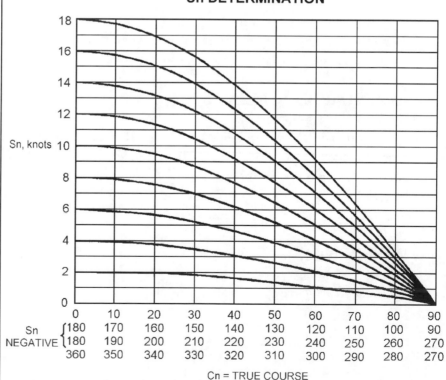

Cn = TRUE COURSE

Figure 1

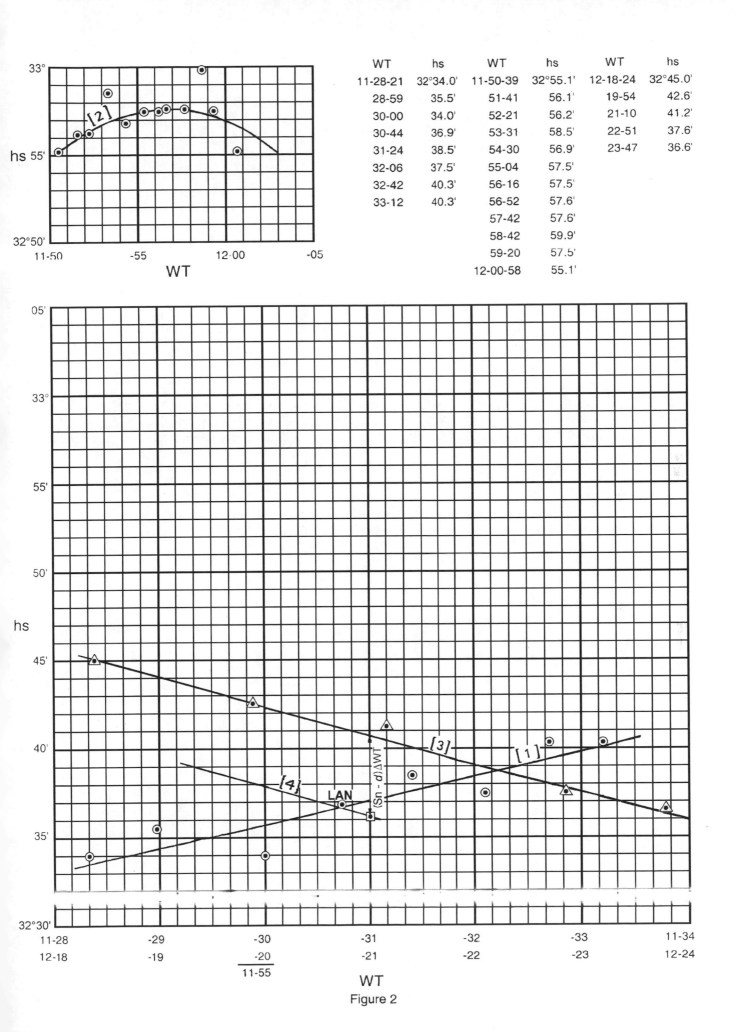

Figure 2

Following are thirty pairs of Meridian Transit Sight Reduction forms and blank sight plots. They can be removed from the book as necessary.

MERIDIAN TRANSIT SIGHT REDUCTION FORM

TIME OF TRANSIT

DR GHA _____°_____.___'

= Lo W, 360° - Lo E

GHA _____ h _____°_____.___'

GHA Diff _____°_____.___'

Min-Sec _____

UT _____

ZD E- ( ) _____   (rev)
   W+

ZT _____

- - - - - - - - - - - - - - - - - -

Last sight WT _____

Difference _____

ZT + Difference _____

LAN

S _____.___ kt   Cn _____°

Sn ( ) _____.___ kt

Sn - d ( ) _____.___'/hr

d positive if change is Northerly

ΔWT _____ m

Δ hs ( ) _____.___'

WT _____

WE f- ( ) _____
    s+

ZT _____

ZD E- ( ) _____
   W+

UT _____

LONGITUDE

GHA _____ h _____°_____.___'

_____ m _____ s _____°_____.___'

_____°_____.___'

360°   359° 59 .10'

Lo _____°_____.___' E

BASIC DATA

Date_____ Body _____

DR L _____°_____.___' N S E W

DR Lo _____°_____.___' N S E W

DECLINATION

Dec_____ hr _____°_____.___' N S

d ( ) _____.___'/hr

  d corr ( ) _____.___'

Dec _____°_____.___' N S

LATITUDE section with scale 50°N to 50°S, EQ marked:

Co-Alt = 90° - Ho _____°_____.___' N S

Co-Alt name is observer position relative to body. Dec _____°_____.___' N S

L = Co-Alt ± Dec _____°_____.___' N S

If Co-Alt and Dec opposite names, subtract.

ALTITUDE

Ht of eye _____.___ ft

hs _____°_____.___'

IC ( ) _____.___'

Dip (−) _____.___'

Total ( ) _____.___'

ha _____°_____.___'

Corr ( ) _____.___'

Ho _____°_____.___'

LATITUDE   90°   89 ° 59 .10'

Sn DETERMINATION

Sn	180	170	160	150	140	130	120	110	100	90
NEGATIVE	180	190	200	210	220	230	240	250	260	270
	360	350	340	330	320	310	300	290	280	270

Cn = TRUE COURSE

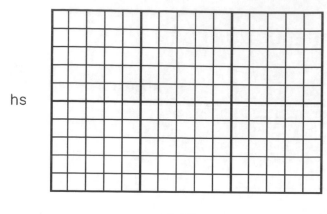

hs

WT

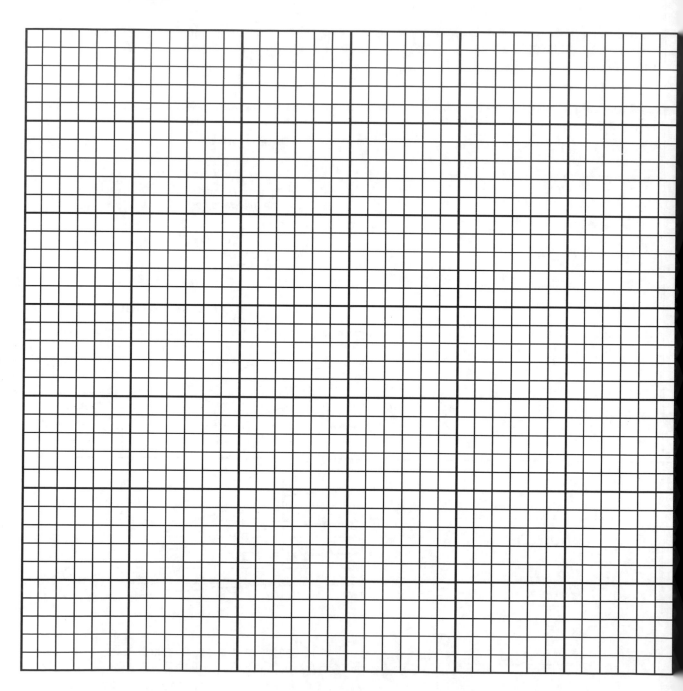

hs

WT

MERIDIAN TRANSIT SIGHT REDUCTION FORM

TIME OF TRANSIT

DR GHA _____ ° _____ . _____ '

= Lo W, 360° - Lo E

GHA _____ h _____ ° _____ . _____ '

GHA Diff _____ ° _____ . _____

Min-Sec _____

UT _____

ZD $\begin{smallmatrix} E- \\ W+ \end{smallmatrix}$ ( ) _____   (rev)

ZT _____

- - - - - - - - - - - - - - - - - - - -

Last sight WT _____

Difference _____

ZT + Difference _____

LAN

S _____ . ___ kt   Cn _____ °

Sn ( ) _____ . ___ kt

Sn - d ( ) _____ . ___ '/hr

d positive if change is Northerly

ΔWT _____ m

Δ hs ( ) _____ . ___ '

WT _____

WE $\begin{smallmatrix} f- \\ s+ \end{smallmatrix}$ ( ) _____

ZT _____

ZD $\begin{smallmatrix} E- \\ W+ \end{smallmatrix}$ ( ) _____

UT _____

LONGITUDE

GHA _____ h _____ ° _____ . _____ '

_____ m _____ s _____ ° _____ . _____ '

360° _____   359° 59 . 10 '

Lo _____ ° _____ . _____ ' E

BASIC DATA

Date _____   Body _____

DR L _____ ° _____ . _____ ' $\begin{smallmatrix} N \\ S \end{smallmatrix}$

DR Lo _____ ° _____ . _____ ' $\begin{smallmatrix} E \\ W \end{smallmatrix}$

DECLINATION

Dec _____ hr _____ ° _____ . _____ ' $\begin{smallmatrix} N \\ S \end{smallmatrix}$

d ( ) _____ . _____ '/hr

d corr ( ) _____ . _____ '

Dec _____ ° _____ . _____ ' $\begin{smallmatrix} N \\ S \end{smallmatrix}$

LATITUDE

$$90° \quad 89 ° \quad 59 . 10 '$$

Co-Alt = 90° - Ho _____ ° _____ . _____ ' $\begin{smallmatrix} N \\ S \end{smallmatrix}$

Co-Alt name is observer
position relative to body.   Dec _____ ° _____ . _____ ' $\begin{smallmatrix} N \\ S \end{smallmatrix}$

L = Co-Alt ± Dec _____ ° _____ . _____ ' $\begin{smallmatrix} N \\ S \end{smallmatrix}$

If Co-Alt and Dec opposite
names, subtract.

ALTITUDE

Ht of eye _____ . _____ ft

hs _____ ° _____ . _____ '

IC ( ) _____ . _____

Dip (–) _____ . _____

Total ( ) _____ . _____

ha _____ ° _____ . _____

Corr ( ) _____ . _____

Ho _____ ° _____ . _____

Sn DETERMINATION

Sn	180	170	160	150	140	130	120	110	100	90
NEGATIVE {	180	190	200	210	220	230	240	250	260	270
	360	350	340	330	320	310	300	290	280	270

Cn = TRUE COURSE

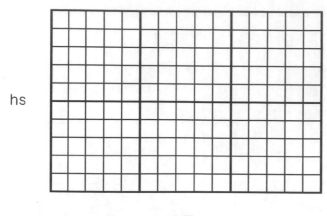

WT

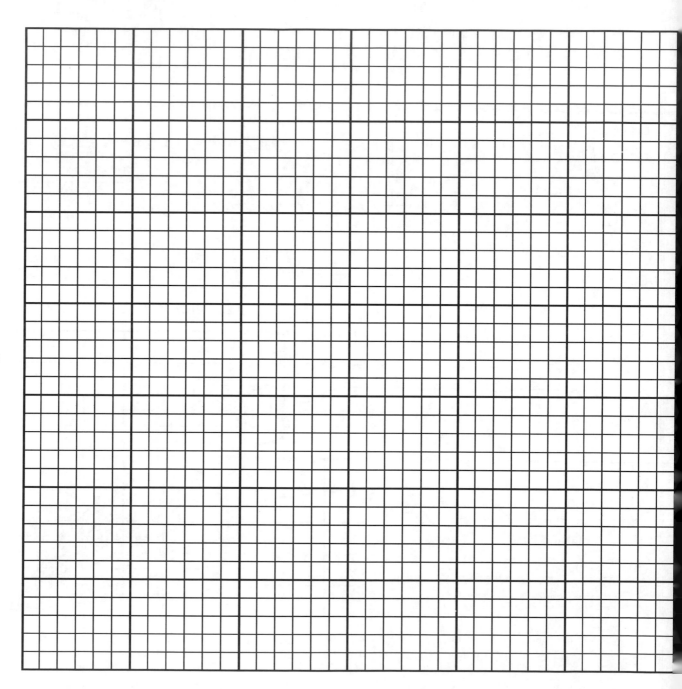

WT

MERIDIAN TRANSIT SIGHT REDUCTION FORM

TIME OF TRANSIT

DR GHA _____°_____.___'

= Lo W, 360° - Lo E

GHA_____ h_____°_____.___'

GHA Diff _____°_____.___'

Min-Sec _____

UT _____

ZD E-／W+ ( )_____ (rev)

ZT _____

- -

Last sight WT _____

Difference _____

ZT + Difference _____

LAN

S _____.___ kt   Cn _____°

Sn ( ) _____.___ kt

Sn - d ( ) _____.___'/hr

d positive if change is Northerly

ΔWT _____ m

Δ hs ( )_____.___'

WT _____

WE f-／s+ ( ) _____

ZT _____

ZD E-／W+ ( )_____

UT _____

LONGITUDE

GHA _____ h _____°_____.___'

_____ m _____ s _____°_____.___'

360° _____359°_59_.10'

Lo _____°_____.___' E

BASIC DATA

Date_____ Body _____

DR L _____°_____.___' N／S／E

DR Lo _____°_____.___' W

DECLINATION

Dec_____ hr _____°_____.___' N／S

d ( )_____.___'/hr

  d corr ( ) _____.___'

Dec _____°_____.___' N／S

ALTITUDE

Ht of eye _____.___ ft

hs _____°_____.___'

  IC ( ) _____.___'

  Dip (−) _____.___'

  Total ( ) _____.___'

ha _____°_____.___'

  Corr ( ) _____.___'

Ho _____°_____.___'

LATITUDE

50°N ⊤

EQ ⊦

50°S ⊥

90°   89° 59 .10'

Co-Alt = 90° - Ho _____°_____.___ N／S

Co-Alt name is observer position relative to body.   Dec _____°_____.___' N／S

L = Co-Alt ± Dec _____°_____.___' N／S

If Co-Alt and Dec opposite names, subtract.

Sn DETERMINATION

Sn NEGATIVE {	180	170	160	150	140	130	120	110	100	90
	180	190	200	210	220	230	240	250	260	270
	360	350	340	330	320	310	300	290	280	270

Cn = TRUE COURSE

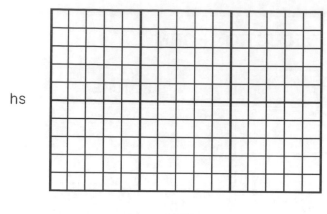

hs

WT

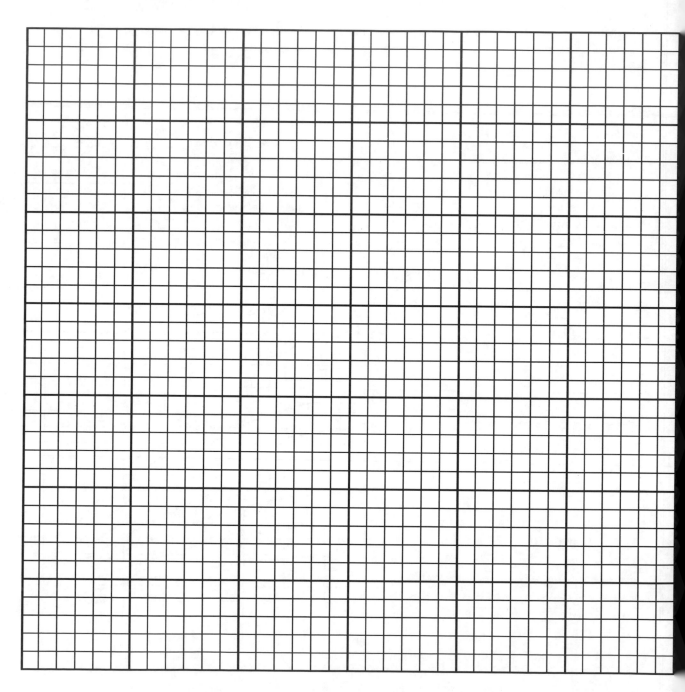

hs

WT

MERIDIAN TRANSIT SIGHT REDUCTION FORM

TIME OF TRANSIT

DR GHA _____ ° _____ . ___ '

= Lo W, 360° - Lo E

GHA ____ h _____ ° ____ . ___ '

GHA Diff _____ ° ____ . ___ '

Min-Sec _____

UT _____

ZD $^{E-}_{W+}$ ( ) _____ (rev)

ZT _____

- - - - - - - - - - - - - - - - - - - -

Last sight WT _____

Difference _____

ZT + Difference _____

LAN

S _____ . ___ kt   Cn _____ °

Sn ( ) _____ . ___ kt

Sn - d ( ) _____ . ___ '/hr

d positive if change is Northerly

ΔWT _____ m

Δ hs ( ) _____ . ___ '

WT _____

WE $^{f-}_{s+}$ ( ) _____

ZT _____

ZD $^{E-}_{W+}$ ( ) _____

UT _____

LONGITUDE

GHA _____ h _____ ° ____ . ___ '

___ m ___ s _____ ° ____ . ___ '

_____ _____ ° ____ . ___ '

360° 359° 59 . 10 '

Lo _____ ° ____ . ___ ' E

BASIC DATA

Date _____ Body _____

DR L _____ ° ____ . ___ ' N/S

DR Lo _____ ° ____ . ___ ' N E S W

DECLINATION

Dec ____ hr _____ ° ____ . ___ ' N S

d ( ) _____ . ___ '/hr

d corr ( ) _____ . ___ '

Dec _____ ° ____ . ___ ' N S

LATITUDE

Co-Alt = 90° - Ho _____ ° ____ . ___ ' N S

Co-Alt name is observer position relative to body.

Dec _____ ° ____ . ___ ' N S

L = Co-Alt ± Dec _____ ° ____ . ___ ' N S

If Co-Alt and Dec opposite names, subtract.

90° 89 ° 59 . 10 '

ALTITUDE

Ht of eye _____ . ___ ft

hs _____ ° ____ . ___ '

IC ( ) _____ . ___

Dip (−) _____ . ___

Total ( ) _____ . ___

ha _____ ° ____ . ___ '

Corr ( ) _____ . ___ '

Ho _____ ° ____ . ___ '

Sn DETERMINATION

Sn	180	170	160	150	140	130	120	110	100	90
NEGATIVE	180	190	200	210	220	230	240	250	260	270
	360	350	340	330	320	310	300	290	280	270

Cn = TRUE COURSE

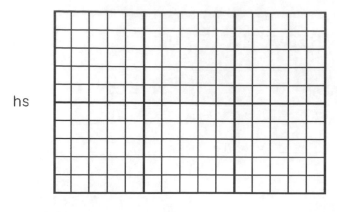

hs

WT

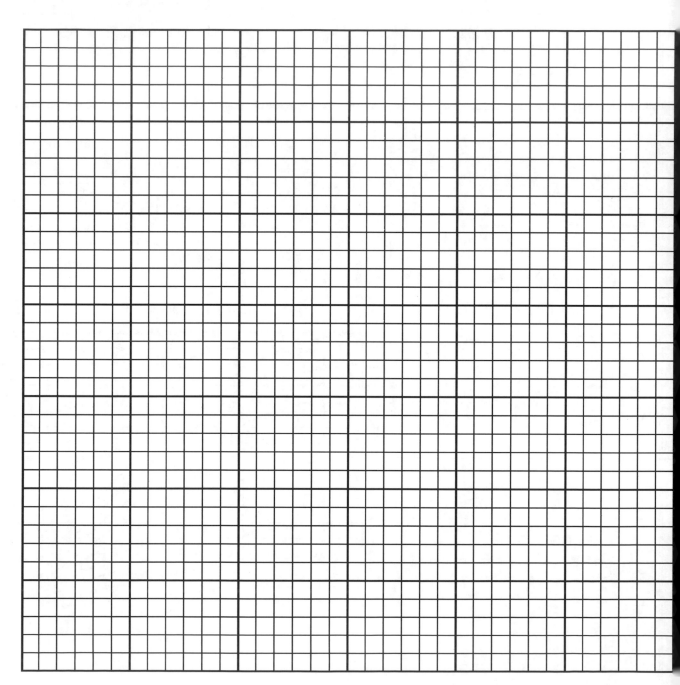

hs

WT

MERIDIAN TRANSIT SIGHT REDUCTION FORM

TIME OF TRANSIT

DR GHA _____°_____.___'

= Lo W, 360° - Lo E

GHA_____ h _____°_____.___'

GHA Diff _____°_____.___'

Min-Sec _____

UT _____

ZD E- ( ) _____   (rev)
   W+

ZT _____

- - - - - - - - - - - - - - - - - - - -

Last sight WT _____

Difference _____

ZT + Difference _____

LAN

S _____.___ kt   Cn _____°

Sn ( ) _____.__ kt

Sn - d ( ) _____.___'/hr

d positive if change is Northerly

ΔWT _____ m

Δ hs ( )_____.___'

WT _____

WE f- ( ) _____
   s+

ZT _____

ZD E- ( )_____
   W+

UT _____

LONGITUDE

GHA _____ h _____°____.___'

___ m ___ s _____°____.___'

Lo _____.___'

360° _____ 359° 59 .10'

Lo _____°____.___' E

BASIC DATA

Date_____ Body _____

DR L _____°_____.___' N/S

DR Lo _____°_____.___' E/W

DECLINATION

Dec_____ hr _____°____.___' N/S

d ( )_____.___'/hr

d corr ( ) _____.___'

Dec _____°____.___' N/S

LATITUDE

90°   89 ° 59 .10'

Co-Alt = 90° - Ho _____°____.___' N/S

Co-Alt name is observer position relative to body.

Dec _____°____.___' N/S

L = Co-Alt ± Dec _____°____.___' N/S

If Co-Alt and Dec opposite names, subtract.

ALTITUDE

Ht of eye _____.___ ft

hs _____°____.___'

IC ( ) _____.___

Dip (−) _____.___

Total ( ) _____.___

ha _____°____.___'

Corr ( ) _____.___

Ho _____°____.___'

Sn DETERMINATION

Sn	180	170	160	150	140	130	120	110	100	90
NEGATIVE	180	190	200	210	220	230	240	250	260	270
	360	350	340	330	320	310	300	290	280	270

Cn = TRUE COURSE

hs

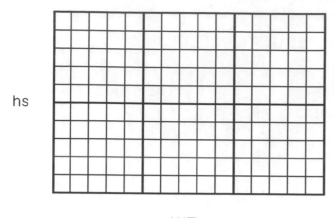

WT

hs

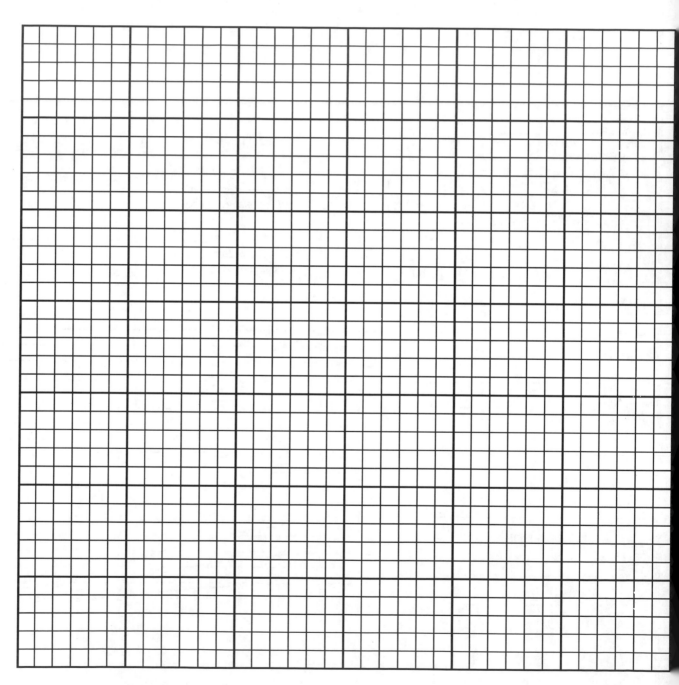

WT

MERIDIAN TRANSIT SIGHT REDUCTION FORM

TIME OF TRANSIT

DR GHA _____°_____.___'

= Lo W, 360° - Lo E

GHA ___ h _____°____.___'

GHA Diff _____°___.___'

Min-Sec _____

UT _____

ZD $^{E-}_{W+}$ ( ) _____ (rev)

ZT _____

- - - - - - - - - - - - - - - - - -

Last sight WT _____

Difference _____

ZT + Difference _____

LAN

S _____.___ kt   Cn _____°

Sn ( ) _____.___ kt

Sn - d ( ) _____.___'/hr

d positive if change is Northerly

ΔWT _____ m

Δ hs ( ) _____.___'

WT _____

WE $^{f-}_{s+}$ ( ) _____

ZT _____

ZD $^{E-}_{W+}$ ( ) _____

UT _____

LONGITUDE

GHA ___ h _____°___.___'

___ m ___ s _____°___.___'

360° _____359° 59 .10'

Lo _____°___.___' E

BASIC DATA

Date_____ Body _____

DR L _____°____.___' ^N_S

DR Lo _____°____.___' ^E_W

DECLINATION

Dec ___ hr _____°____.___' ^N_S

d ( ) ___.___'/hr

d corr ( ) _____.___'

Dec _____°____.___' ^N_S

LATITUDE

_____

50°N

EQ

50°S

$90°$ _____89° 59 .10'

Co-Alt = 90° - Ho _____°____.___' ^N_S

Co-Alt name is observer position relative to body.

Dec _____°____.___' ^N_S

L = Co-Alt ± Dec _____°____.___' ^N_S

If Co-Alt and Dec opposite names, subtract.

ALTITUDE

Ht of eye _____.___ ft

hs _____°____.___'

IC ( ) _____.___'

Dip (−) _____.___'

Total ( ) _____.___'

ha _____°____.___'

Corr ( ) _____.___'

Ho _____°____.___'

Sn DETERMINATION

Sn {	180	170	160	150	140	130	120	110	100	90
NEGATIVE {	180	190	200	210	220	230	240	250	260	270
	360	350	340	330	320	310	300	290	280	270

Cn = TRUE COURSE

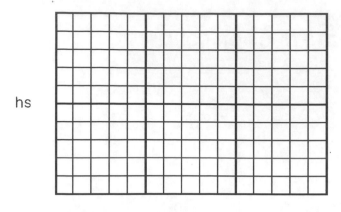

hs

WT

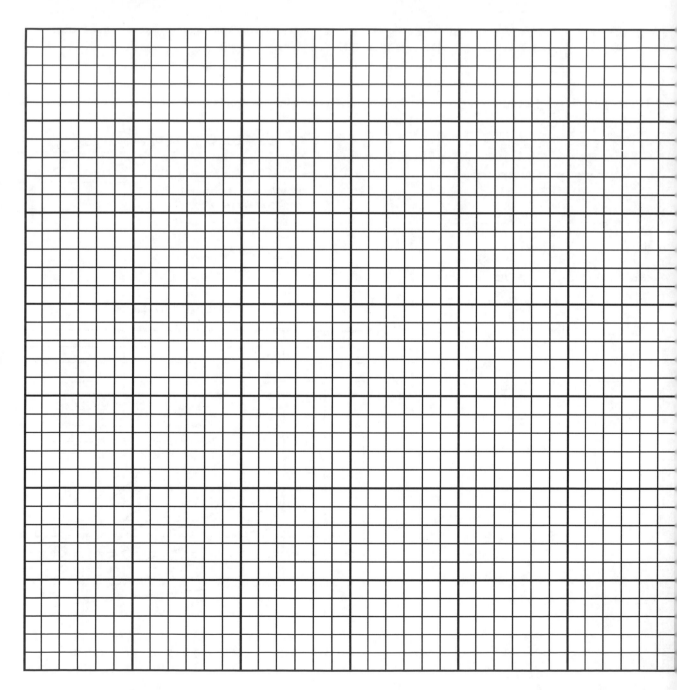

hs

WT

MERIDIAN TRANSIT SIGHT REDUCTION FORM

TIME OF TRANSIT

DR GHA _____°_____.____'

= Lo W, 360° - Lo E

GHA _____ h _____°_____.____'

GHA Diff _____°____.____'

Min-Sec _____

UT _____

ZD $^{E-}_{W+}$ ( )_____ (rev)

ZT _____

- -

Last sight WT _____

Difference _____

ZT + Difference _____

LAN

S _____.___ kt   Cn _____°

Sn ( ) _____.___ kt

Sn - d ( ) _____.___'/hr

d positive if change is Northerly

ΔWT _____ m

Δ hs ( )_____.___'

WT _____

WE $^{f-}_{s+}$ ( ) _____

ZT _____

ZD $^{E-}_{W+}$ ( )_____

UT _____

LONGITUDE

GHA _____ h _____°_____.____'

____ m ___ s _____°_____.____'

360° _____359°_59_.10'

Lo _____°_____.____' E

BASIC DATA

Date_____ Body _____

DR L _____°____.____' N/S

DR Lo _____°____.____' N/E/S/W

DECLINATION

Dec____ hr _____°____.____' N/S

d ( )____.____'/hr

d corr ( ) _____.____'

Dec _____°____.____' N/S

LATITUDE

90° ____ 89 ° 59 . 10 '

Co-Alt = 90° - Ho _____°____.____' N/S

Co-Alt name is observer position relative to body.

Dec _____°____.____' N/S

L = Co-Alt ± Dec _____°____.____' N/S

If Co-Alt and Dec opposite names, subtract.

(Latitude scale: 50°N — EQ — 50°S)

ALTITUDE

Ht of eye _____.___ ft

hs _____°____.____'

IC ( ) _____.____'

Dip (−) _____.____'

Total ( ) _____.____'

ha _____°____.____'

Corr ( ) _____.____'

Ho _____°____.____'

Sn DETERMINATION

Sn, knots (vertical axis: 0, 2, 4, 6, 8, 10, 12, 14, 16, 18)

Sn POSITIVE	180	170	160	150	140	130	120	110	100	90
Sn NEGATIVE	180	190	200	210	220	230	240	250	260	270
	360	350	340	330	320	310	300	290	280	270

Cn = TRUE COURSE

hs

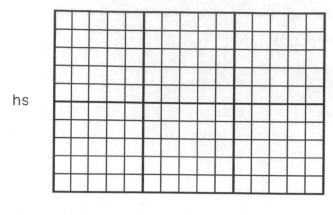

WT

hs

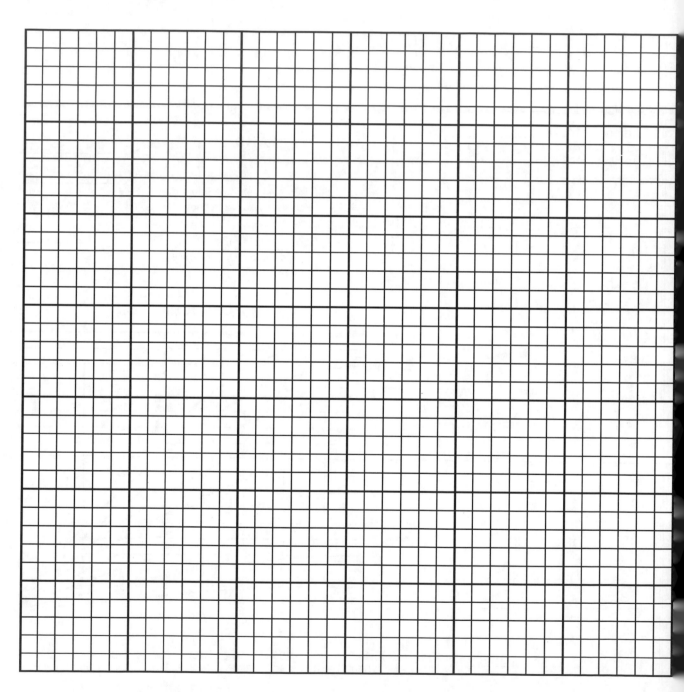

WT

MERIDIAN TRANSIT SIGHT REDUCTION FORM

TIME OF TRANSIT

DR GHA _____ ° _____ . ___'

= Lo W, 360° - Lo E

GHA _____ h _____ ° _____ . ___'

GHA Diff _____ ° _____ . ___'

Min-Sec _____

UT _____

ZD E- ( ) _____ (rev)
W+

ZT _____

- -

Last sight WT _____

Difference _____

ZT + Difference _____

LAN

S _____ . ___ kt   Cn _____ °

Sn ( ) _____ . ___ kt

Sn - d ( ) _____ . ___'/hr

d positive if change is Northerly

ΔWT _____ m

Δ hs ( ) _____ . ___'

WT _____

WE f- ( ) _____
   s+

ZT _____

ZD E- ( ) _____
   W+

UT _____

LONGITUDE

GHA _____ h _____ ° _____ . ___'

_____ m _____ s _____ ° _____ . ___'

_____

360° _____ 359° 59 . 10'

Lo _____ ° _____ . ___' E

BASIC DATA

Date_____ Body _____

DR L _____ ° _____ . ___' N
                             S
                             E
DR Lo _____ ° _____ . ___' W

DECLINATION

Dec _____ hr _____ ° _____ . ___' N
                                    S

d ( ) _____ . ___'/hr

d corr ( ) _____ . ___'

Dec _____ ° _____ . ___' N
                              S

ALTITUDE

Ht of eye _____ . ___ ft

hs _____ ° _____ . ___'

IC ( ) _____ . ___'

Dip (−) _____ . ___'

Total ( ) _____ . ___'

ha _____ ° _____ . ___'

Corr ( ) _____ . ___'

Ho _____ ° _____ . ___'

LATITUDE

50°N —

EQ —

50°S —

90° _____ 89 ° 59 . 10'

Co-Alt = 90° - Ho _____ ° _____ . ___' N
                                         S
Co-Alt name is observer   N
position relative to body. Dec _____ ° _____ . ___' S

                                         N
L = Co-Alt ± Dec _____ ° _____ . ___' S
If Co-Alt and Dec opposite
names, subtract.

Sn DETERMINATION

Sn	180	170	160	150	140	130	120	110	100	90
NEGATIVE	180	190	200	210	220	230	240	250	260	270
	360	350	340	330	320	310	300	290	280	270

Cn = TRUE COURSE

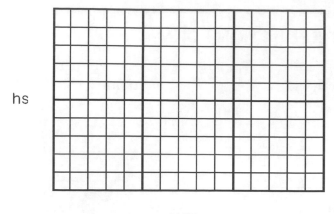

WT

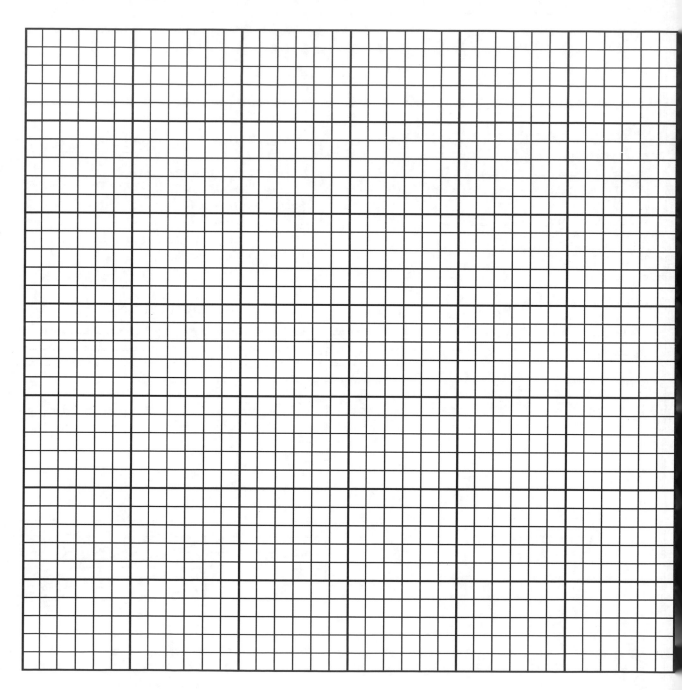

WT

MERIDIAN TRANSIT SIGHT REDUCTION FORM

TIME OF TRANSIT

DR GHA _____°_____.___'

  = *Lo W, 360° - Lo E*

GHA_____h_____°_____.___'

GHA Diff _____°_____.___'

Min-Sec _____

UT _____

ZD $^{E-}_{W+}$ ( )_____ (rev)

ZT _____

- - - - - - - - - - - - - - - - - - - -

Last sight WT _____

Difference _____

ZT + Difference _____

LAN

S _____.___ kt   Cn _____°

Sn ( ) _____.___ kt

Sn - *d* ( ) _____.___'/hr

  d positive if change is Northerly

ΔWT _____ m

Δ hs ( ) _____.___'

WT _____

WE $^{f-}_{s+}$ ( ) _____

ZT _____

ZD $^{E-}_{W+}$ ( )_____

UT _____

LONGITUDE

GHA _____h _____°_____.___'

_____m _____s _____°_____.___'

Lo _____

360°   359° 59 .10'

Lo _____°_____.___' E

BASIC DATA

Date_____ Body _____

DR L _____°_____.___' $^{N}_{S}$

DR Lo _____°_____.___' $^{E}_{W}$

DECLINATION

Dec_____hr_____°_____.___' $^{N}_{S}$

d ( )_____.___'/hr

 d corr ( ) _____.___'

Dec _____°_____.___' $^{N}_{S}$

LATITUDE

90°   89 ° 59 . 10 '

Co-Alt = 90° - Ho _____°_____.___' $^{N}_{S}$

*Co-Alt name is observer
position relative to body.*

Dec _____°_____.___' $^{N}_{S}$

L = Co-Alt ± Dec _____°_____.___' $^{N}_{S}$

*If Co-Alt and Dec opposite
names, subtract.*

ALTITUDE

Ht of eye _____.___ ft

hs _____°_____.___'

IC ( ) _____.___'

Dip (−) _____.___'

Total ( ) _____.___'

ha _____°_____.___'

Corr ( ) _____.___'

Ho _____°_____.___'

Sn DETERMINATION

Sn $\{$	180	170	160	150	140	130	120	110	100	90
NEGATIVE $\{$	180	190	200	210	220	230	240	250	260	270
	360	350	340	330	320	310	300	290	280	270

Cn = TRUE COURSE

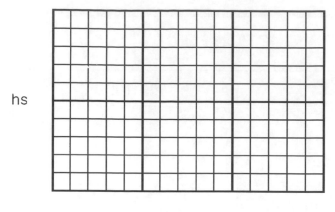

WT

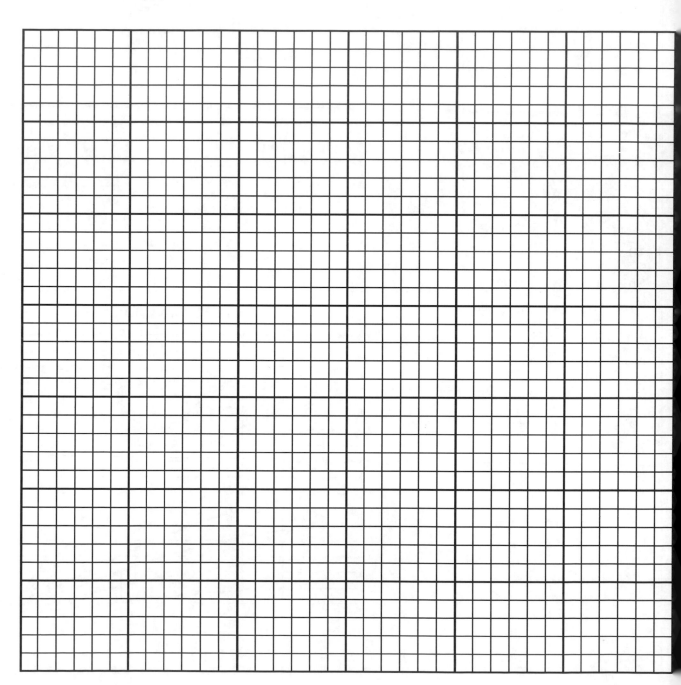

WT

hs

MERIDIAN TRANSIT SIGHT REDUCTION FORM

TIME OF TRANSIT

DR GHA _____°____.___'

= Lo W, 360° - Lo E

GHA ____ h _____°___.__'

GHA Diff _____°___.__'

Min-Sec _____

UT _____

ZD $^{E-}_{W+}$ ( ) _____ (rev)

ZT _____

- - - - - - - - - - - - - - - - - -

Last sight WT _____

Difference _____

ZT + Difference _____

LAN

S ____.__ kt   Cn _____°

Sn ( ) _____.__ kt

Sn - d ( ) _____.__'/hr

d positive if change is Northerly

ΔWT _____ m

Δ hs ( ) _____.__'

WT _____

WE $^{f-}_{s+}$ ( ) _____

ZT _____

ZD $^{E-}_{W+}$ ( ) _____

UT _____

LONGITUDE

GHA ____ h ____°___.__'

___ m ___ s ___°___.__'

Lo _____ ___°___.__'

360°   359° 59 .10 '

Lo _____°___.__' E

BASIC DATA

Date_____ Body _____

DR L _____°___.__ ^N_S

DR Lo _____°___.__ $^{E}_{W}$

DECLINATION

Dec____ hr ____°___.__' ^N_S

d ( ) ____.__'/hr

   d corr ( ) _____.__'

Dec _____°___.__' ^N_S

LATITUDE

90°   89 ° 59 .10 '

Co-Alt = 90° - Ho _____°___.__' ^N_S

Co-Alt name is observer position relative to body.

Dec _____°___.__' ^N_S

L = Co-Alt ± Dec _____°___.__' ^N_S

If Co-Alt and Dec opposite names, subtract.

ALTITUDE

Ht of eye ____.__ ft

hs _____°___.__

   IC ( ) ____.__

   Dip (−) ____.__

   Total ( ) ____.__'

ha _____°___.__

   Corr ( ) ____.__'

Ho _____°___.__

Sn DETERMINATION

Sn		180	170	160	150	140	130	120	110	100	90
NEGATIVE	{	180	190	200	210	220	230	240	250	260	270
		360	350	340	330	320	310	300	290	280	270

Cn = TRUE COURSE

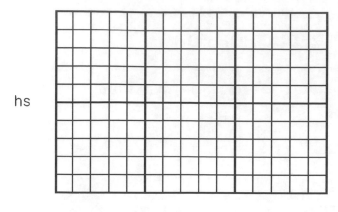

hs

WT

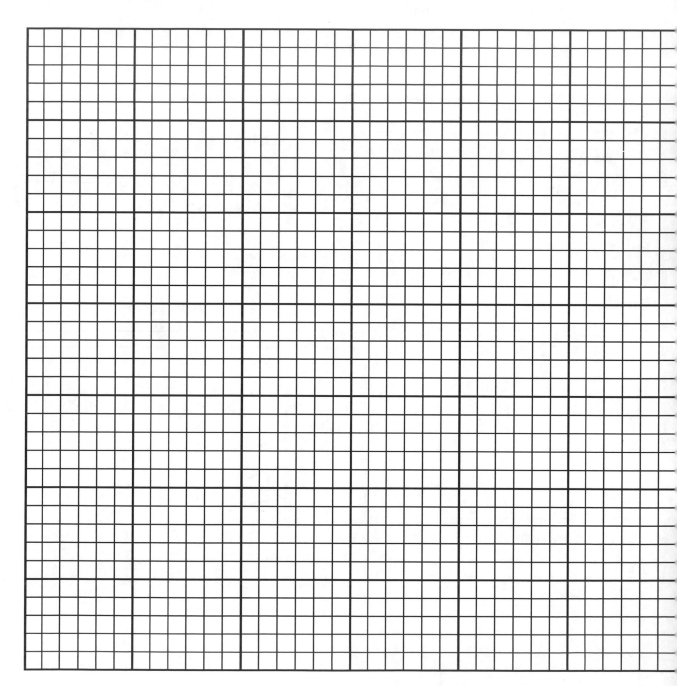

hs

WT

MERIDIAN TRANSIT SIGHT REDUCTION FORM

TIME OF TRANSIT

DR GHA _____°_____.___'

= *Lo W, 360° - Lo E*

GHA_____h_____°_____.___'

GHA Diff _____°_____.___'

Min-Sec _____

UT _____

ZD $_{W+}^{E-}$ ( )_____ (rev)

ZT _____

- - - - - - - - - - - - - - - - - - - -

Last sight WT _____

Difference _____

ZT + Difference _____

LAN

S _____.___kt   Cn _____°

Sn ( ) _____.___kt

Sn - *d* ( ) _____.___'/hr

d positive if change is Northerly

ΔWT _____ m

Δ hs ( )_____.___'

WT _____

WE $_{s+}^{f-}$ ( ) _____

ZT _____

ZD $_{W+}^{E-}$ ( )_____

UT _____

LONGITUDE

GHA _____h _____°_____.___'

_____m _____s _____°_____.___'

360° _____359°_59_.10_'

Lo _____°_____.___' E

BASIC DATA

Date_____ Body _____

DR L _____°_____.___' $_{S}^{N}$

DR Lo _____°_____.___' $_{W}^{S E}$

DECLINATION

Dec_____hr _____°_____.___' $_{S}^{N}$

d ( )_____.___'/hr

d corr ( ) _____.___'

Dec _____°_____.___' $_{S}^{N}$

LATITUDE

50°N
EQ
50°S

Co-Alt = 90° - Ho

*Co-Alt name is observer
position relative to body.*

L = Co-Alt ± Dec

*If Co-Alt and Dec opposite
names, subtract.*

90°_____89_°_59_.10_'

_____°_____.___' $_{S}^{N}$

Dec _____°_____.___' $_{S}^{N}$

_____°_____.___' $_{S}^{N}$

ALTITUDE

Ht of eye _____.___ft

hs _____°_____.___'

IC ( ) _____.___'

Dip (−) _____.___'

Total ( ) _____.___'

ha _____°_____.___'

Corr ( ) _____.___'

Ho _____°_____.___'

Sn DETERMINATION

Sn NEGATIVE	180	170	160	150	140	130	120	110	100	90
	180	190	200	210	220	230	240	250	260	270
	360	350	340	330	320	310	300	290	280	270

Cn = TRUE COURSE

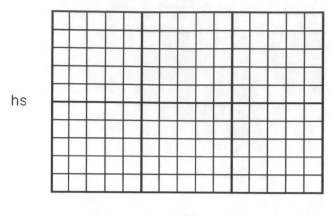

hs

WT

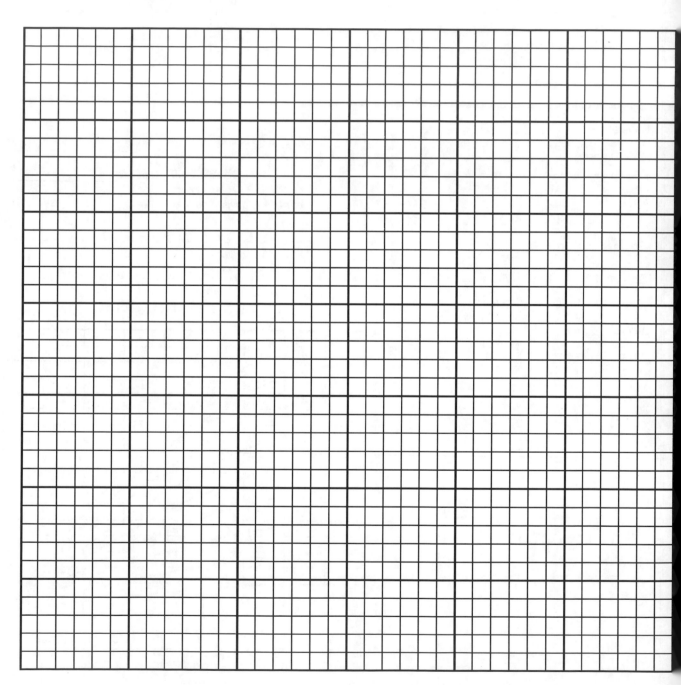

hs

WT

MERIDIAN TRANSIT SIGHT REDUCTION FORM

TIME OF TRANSIT

DR GHA _____ ° ____ . ___ '

= Lo W, 360° - Lo E

GHA ____ h _____ ° ____ . ___ '

GHA Diff _____ ° ____ . ___ '

Min-Sec _____

UT _____

ZD $\begin{smallmatrix}E-\\W+\end{smallmatrix}$ ( ) _____ (rev)

ZT _____

Last sight WT _____

Difference _____

ZT + Difference _____

LAN

S _____ . ___ kt   Cn _____ °

Sn ( ) _____ . ___ kt

Sn - d ( ) _____ . ___ '/hr

d positive if change is Northerly

ΔWT _____ m

Δ hs ( ) _____ . ___ '

WT _____

WE $\begin{smallmatrix}f-\\s+\end{smallmatrix}$ ( ) _____

ZT _____

ZD $\begin{smallmatrix}E-\\W+\end{smallmatrix}$ ( ) _____

UT _____

LONGITUDE

GHA ____ h _____ ° ____ . ___ '

___ m ___ s _____ ° ____ . ___ '

360° _____ 359° 59 . 10 '

Lo _____ ° ____ . ___ ' E

BASIC DATA

Date _____ Body _____

DR L _____ ° ____ . ___ ' $\begin{smallmatrix}N\\S\end{smallmatrix}$

DR Lo _____ ° ____ . ___ ' $\begin{smallmatrix}E\\W\end{smallmatrix}$

DECLINATION

Dec ____ hr _____ ° ____ . ___ ' $\begin{smallmatrix}N\\S\end{smallmatrix}$

d ( ) ____ . ___ '/hr

d corr ( ) _____ . ___ '

Dec _____ ° ____ . ___ ' $\begin{smallmatrix}N\\S\end{smallmatrix}$

LATITUDE

90°   89 ° 59 . 10 '

Co-Alt = 90° - Ho _____ ° ____ . ___ ' $\begin{smallmatrix}N\\S\end{smallmatrix}$

Co-Alt name is observer position relative to body.

Dec _____ ° ____ . ___ ' $\begin{smallmatrix}N\\S\end{smallmatrix}$

L = Co-Alt ± Dec _____ ° ____ . ___ ' $\begin{smallmatrix}N\\S\end{smallmatrix}$

If Co-Alt and Dec opposite names, subtract.

ALTITUDE

Ht of eye _____ . ___ ft

hs _____ ° ____ . ___ '

IC ( ) _____ . ___ '

Dip (−) _____ . ___ '

Total ( ) _____ . ___ '

ha _____ ° ____ . ___ '

Corr ( ) _____ . ___ '

Ho _____ ° ____ . ___ '

Sn DETERMINATION

Sn	180	170	160	150	140	130	120	110	100	90
NEGATIVE	180	190	200	210	220	230	240	250	260	270
	360	350	340	330	320	310	300	290	280	270

Cn = TRUE COURSE

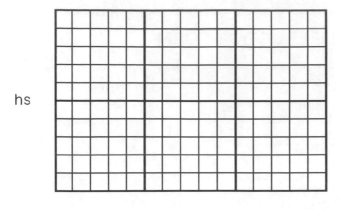

hs

WT

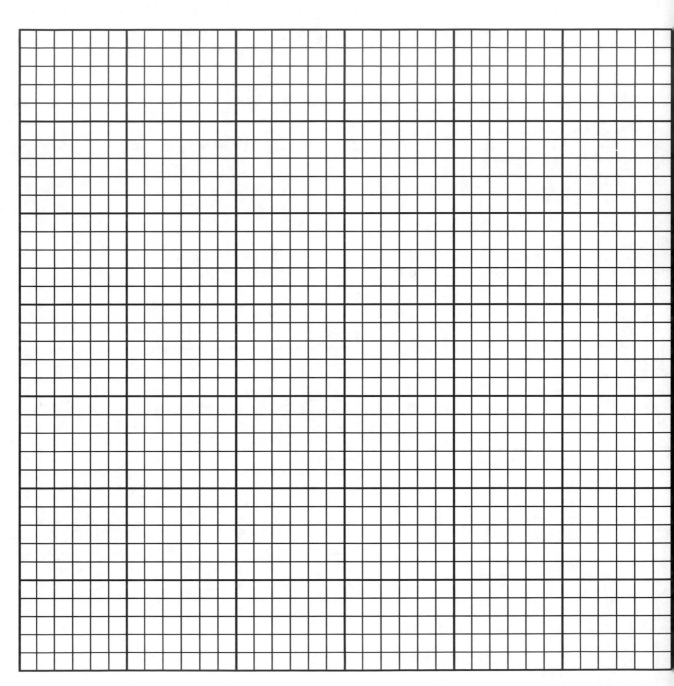

hs

WT

MERIDIAN TRANSIT SIGHT REDUCTION FORM

TIME OF TRANSIT

DR GHA _____°_____.___'

= *Lo W, 360° - Lo E*

GHA ____ h_____°_____.___'

GHA Diff _____°_____.___'

Min-Sec _____

UT _____

ZD $_{W+}^{E-}$ ( ) _____ (rev)

ZT _____

- - - - - - - - - - - - - - - - - - - -

Last sight WT _____

Difference _____

ZT + Difference _____

LAN

S _____.___ kt   Cn _____°

Sn ( ) _____.___ kt

Sn - *d* ( ) _____.___'/hr

d positive if change is Northerly

ΔWT _____ m

Δ hs ( ) _____.___'

WT _____

WE $_{s+}^{f-}$ ( ) _____

ZT _____

ZD $_{W+}^{E-}$ ( ) _____

UT _____

LONGITUDE

GHA ____ h _____°_____.___'

____ m ____ s _____°_____.___'

____ _____

360° 359°59.10'

Lo _____°_____.___' E

BASIC DATA

Date_____ Body _____

DR L _____°_____.___' $_{S}^{N}$

DR Lo _____°_____.___' $_{W}^{E}$

DECLINATION

Dec ____ hr _____°_____.___' $_{S}^{N}$

d ( ) ____.___'/hr

d corr ( ) _____.___'

Dec _____°_____.___' $_{S}^{N}$

ALTITUDE

Ht of eye _____.___ ft

hs _____°_____.___

IC ( ) _____.___

Dip (−) _____.___

Total ( ) _____.___

ha _____°_____.___

Corr ( ) _____.___

Ho _____°_____.___'

LATITUDE

90°   89°59.10'

Co-Alt = 90° - Ho _____°_____.___' $_{S}^{N}$

Co-Alt name is observer position relative to body.   Dec _____°_____.___' $_{S}^{N}$

L = Co-Alt ± Dec _____°_____.___' $_{S}^{N}$

If Co-Alt and Dec opposite names, subtract.

50°N —
EQ —
50°S —

Sn DETERMINATION

Sn	180	170	160	150	140	130	120	110	100	90
NEGATIVE	180	190	200	210	220	230	240	250	260	270
	360	350	340	330	320	310	300	290	280	270

Cn = TRUE COURSE

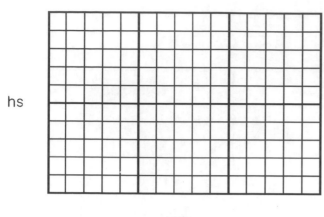

hs

WT

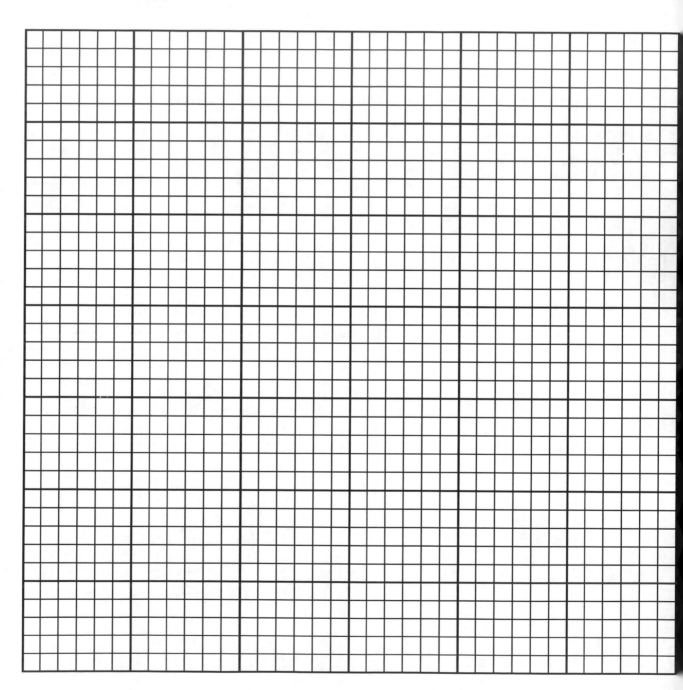

hs

WT

MERIDIAN TRANSIT SIGHT REDUCTION FORM

TIME OF TRANSIT

DR GHA _____°_____.____'

= Lo W, 360° - Lo E

GHA _____ h _____°_____.____'

GHA Diff _____°_____.____'

Min-Sec _____

UT _____

ZD $^{E-}_{W+}$ ( ) _____ (rev)

ZT _____

- - - - - - - - - - - - - - - - - - - -

Last sight WT _____

Difference _____

ZT + Difference _____

LAN

S _____.___ kt   Cn _____°

Sn ( ) _____.___ kt

Sn - d ( ) _____.___'/hr

d positive if change is Northerly

ΔWT _____ m

Δ hs ( ) _____.___'

WT _____

WE $^{f-}_{s+}$ ( ) _____

ZT _____

ZD $^{E-}_{W+}$ ( ) _____

UT _____

LONGITUDE

GHA _____ h _____°_____.____'

_____ m _____ s _____°_____.____'

360° 359° 59 .10 '

Lo _____°_____.____' E

BASIC DATA

Date _____ Body _____

DR L _____°_____.____' $^{N}_{S}$

DR Lo _____°_____.____' $^{E}_{W}$

DECLINATION

Dec _____ hr _____°_____.____' $^{N}_{S}$

d ( ) _____.____'/hr

  d corr ( ) _____.___'

Dec _____°_____.____' $^{N}_{S}$

LATITUDE

50°N

EQ

50°S

90°   89 ° 59 . 10 '

Co-Alt = 90° - Ho _____°_____.____' $^{N}_{S}$

Co-Alt name is observer position relative to body.

Dec _____°_____.____' $^{N}_{S}$

L = Co-Alt ± Dec _____°_____.____' $^{N}_{S}$

If Co-Alt and Dec opposite names, subtract.

ALTITUDE

Ht of eye _____.____ ft

hs _____°_____.____'

IC ( ) _____.___'

Dip (−) _____.___'

Total ( ) _____.___'

ha _____°_____.____'

Corr ( ) _____.___'

Ho _____°_____.____'

Sn DETERMINATION

Sn {	180	170	160	150	140	130	120	110	100	90
NEGATIVE {	180	190	200	210	220	230	240	250	260	270
	360	350	340	330	320	310	300	290	280	270

Cn = TRUE COURSE

hs

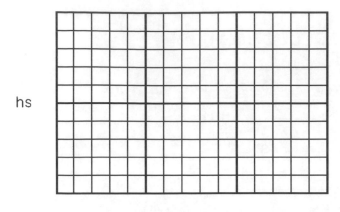

WT

hs

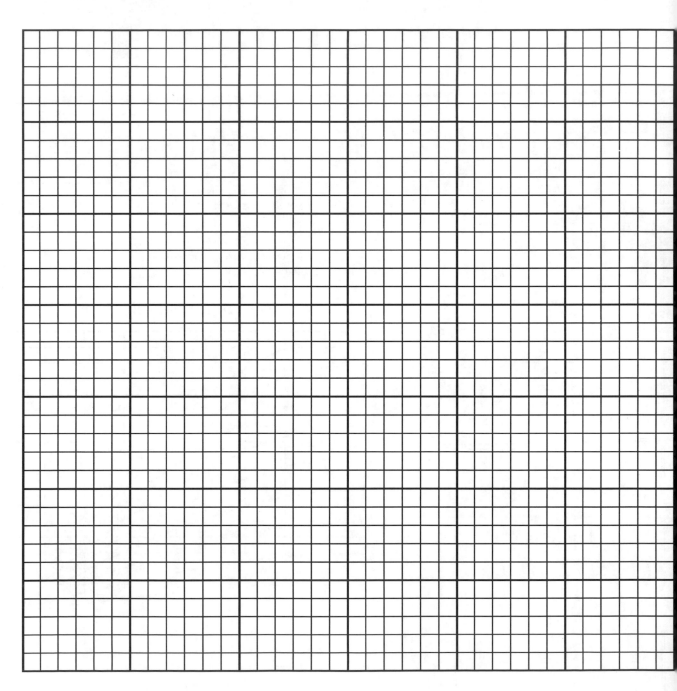

WT

MERIDIAN TRANSIT SIGHT REDUCTION FORM

TIME OF TRANSIT

DR GHA _____°____.___'

= Lo W, 360° - Lo E

GHA ____ h_____°____.___'

GHA Diff _____°____.___'

Min-Sec _____

UT _____

ZD $\begin{matrix} E- \\ W+ \end{matrix}$ ( ) _____ (rev)

ZT _____

- -

Last sight WT _____

Difference _____

ZT + Difference _____

LAN

S ____.___ kt   Cn ____°

Sn ( ) ____.___ kt

Sn - d ( ) ____.___'/hr

d positive if change is Northerly

ΔWT ____ m

Δ hs ( )____.___'

WT _____

WE $\begin{matrix} f- \\ s+ \end{matrix}$ ( ) _____

ZT _____

ZD $\begin{matrix} E- \\ W+ \end{matrix}$ ( ) ____

UT _____

LONGITUDE

GHA ____ h ____°____.___'

____ m ____ s ____°____.___'

360° ____259° 59 .10'

Lo ____°____.___' E

BASIC DATA

Date_____ Body_____

DR L ____°____.___' $\begin{matrix} N \\ S \end{matrix}$

DR Lo ____°____.___' $\begin{matrix} E \\ W \end{matrix}$

DECLINATION

Dec ____ hr ____°____.___' $\begin{matrix} N \\ S \end{matrix}$

d ( )____.___'/hr

d corr ( ) ____.___'

Dec ____°____.___' $\begin{matrix} N \\ S \end{matrix}$

LATITUDE

Co-Alt = 90° - Ho ____°____.___' $\begin{matrix} N \\ S \end{matrix}$

Co-Alt name is observer position relative to body.

Dec ____°____.___' $\begin{matrix} N \\ S \end{matrix}$

L = Co-Alt ± Dec ____°____.___' $\begin{matrix} N \\ S \end{matrix}$

If Co-Alt and Dec opposite names, subtract.

90° 89 ° 59 .10'

(Latitude scale: 50°N — EQ — 50°S)

ALTITUDE

Ht of eye ____.___ ft

hs ____°____.___'

IC ( ) ____.___'

Dip (−) ____.___'

Total ( ) ____.___'

ha ____°____.___'

Corr ( ) ____.___'

Ho ____°____.___'

Sn DETERMINATION

Sn	180	170	160	150	140	130	120	110	100	90
NEGATIVE {	180	190	200	210	220	230	240	250	260	270
	360	350	340	330	320	310	300	290	280	270

Cn = TRUE COURSE

hs

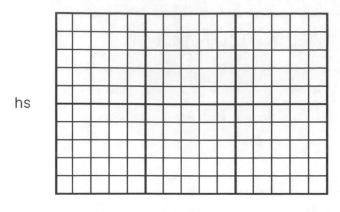

WT

hs

WT

MERIDIAN TRANSIT SIGHT REDUCTION FORM

TIME OF TRANSIT

DR GHA _____ ° _____ . ___ '

   = *Lo W, 360° - Lo E*

GHA _____ h _____ ° _____ . ___ '

GHA Diff _____ ° _____ . ___ '

Min-Sec _____

UT _____

ZD $^{E-}_{W+}$ ( ) _____ (rev)

ZT _____

- -

Last sight WT _____

Difference _____

ZT + Difference _____

LAN

S _____ . ___ kt   Cn _____ °

Sn ( ) _____ . ___ kt

Sn - *d* ( ) _____ . ___ '/hr

  d positive if change is Northerly

ΔWT _____ m

Δ hs ( ) _____ . ___ '

WT _____

WE $^{f-}_{s+}$ ( ) _____

ZT _____

ZD $^{E-}_{W+}$ ( ) _____

UT _____

LONGITUDE

GHA _____ h _____ ° _____ . ___ '

____ m ____ s _____ ° _____ . ___ '

360°   359° 59 . 10 '

Lo _____ ° _____ . ___ ' E

BASIC DATA

Date _____ Body _____

DR L _____ ° _____ . ___ ' $^{N}_{S}$

DR Lo _____ ° _____ . ___ ' $^{E}_{W}$

DECLINATION

Dec ____ hr _____ ° _____ . ___ ' $^{N}_{S}$

d ( ) _____ . ___ '/hr

  d corr ( ) _____ . ___ '

Dec _____ ° _____ . ___ ' $^{N}_{S}$

LATITUDE

90°   89 ° 59 . 10 '

Co-Alt = 90° - Ho _____ ° _____ . ___ ' $^{N}_{S}$

Co-Alt name is observer position relative to body.   Dec _____ ° _____ . ___ ' $^{N}_{S}$

L = Co-Alt ± Dec _____ ° _____ . ___ ' $^{N}_{S}$

If Co-Alt and Dec opposite names, subtract.

(latitude scale: 50°N — EQ — 50°S)

ALTITUDE

Ht of eye _____ . ___ ft

hs _____ ° _____ . ___ '

IC ( ) _____ . ___ '

Dip (−) _____ . ___ '

Total ( ) _____ . ___ '

ha _____ ° _____ . ___ '

Corr ( ) _____ . ___ '

Ho _____ ° _____ . ___ '

Sn DETERMINATION

Sn	180	170	160	150	140	130	120	110	100	90
NEGATIVE	180	190	200	210	220	230	240	250	260	270
	360	350	340	330	320	310	300	290	280	270

Cn = TRUE COURSE

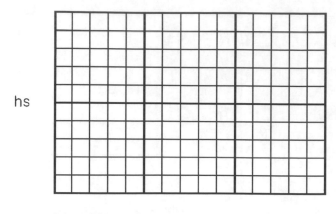

hs

WT

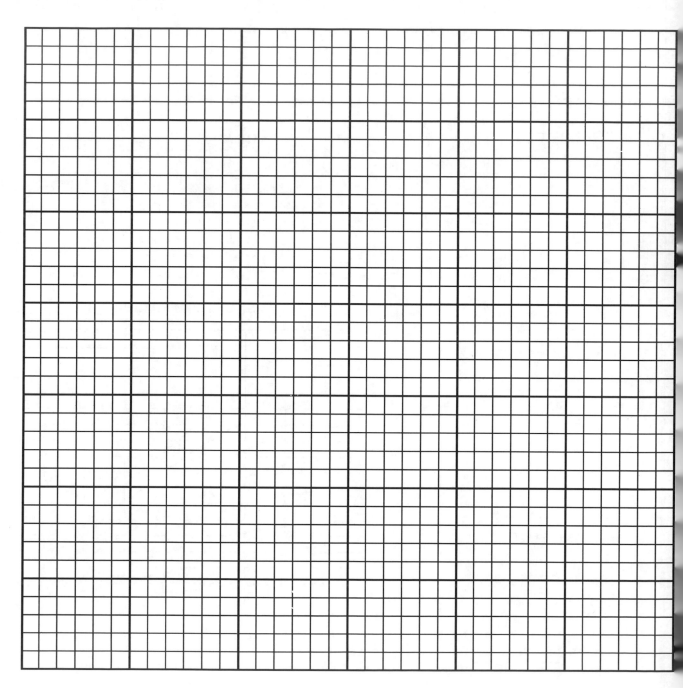

hs

WT

MERIDIAN TRANSIT SIGHT REDUCTION FORM

TIME OF TRANSIT

DR GHA _____ ° _____ . ____ '

= Lo W, 360° - Lo E

GHA _____ h _____ ° ___ . ___ '

GHA Diff _____ ° ___ . ___ '

Min-Sec _____

UT _____

ZD $^{E-}_{W+}$ ( ) _____ (rev)

ZT _____

- -

Last sight WT _____

Difference _____

ZT + Difference _____

LAN

S _____ . ___ kt   Cn _____ °

Sn ( ) _____ . ___ kt

Sn - d ( ) _____ . ___ '/hr

d positive if change is Northerly

ΔWT _____ m

Δ hs ( ) _____ . ___ '

WT _____

WE $^{f-}_{s+}$ ( ) _____

ZT _____

ZD $^{E-}_{W+}$ ( ) _____

UT _____

LONGITUDE

GHA _____ h _____ ° ___ . ___ '

___ m ___ s _____ ° ___ . ___ '

360° _____ 359° 59 .10 '

Lo _____ ° ___ . ___ ' E

BASIC DATA

Date _____ Body _____

DR L _____ ° ___ . ___ ' ^N_S

DR Lo _____ ° ___ . ___ ' ^E_W

DECLINATION

Dec _____ hr _____ ° ___ . ___ ' ^N_S

d ( ) ___ . ___ '/hr

d corr ( ) _____ . ___ '

Dec _____ ° ___ . ___ ' ^N_S

LATITUDE

50°N ─┬
      │
EQ ──┤
      │
50°S ─┴

90° _____ 89 ° 59 . 10 '

Co-Alt = 90° - Ho _____ ° ___ . ___ ' ^N_S

Co-Alt name is observer position relative to body.

Dec _____ ° ___ . ___ ' ^N_S

L = Co-Alt ± Dec _____ ° ___ . ___ ' ^N_S

If Co-Alt and Dec opposite names, subtract.

ALTITUDE

Ht of eye _____ . ___ ft

hs _____ ° ___ . ___ '

IC ( ) _____ . ___ '

Dip (−) _____ . ___ '

Total ( ) _____ . ___ '

ha _____ ° ___ . ___ '

Corr ( ) _____ . ___ '

Ho _____ ° ___ . ___ '

Sn DETERMINATION

Sn	180	170	160	150	140	130	120	110	100	90
NEGATIVE	180	190	200	210	220	230	240	250	260	270
	360	350	340	330	320	310	300	290	280	270

Cn = TRUE COURSE

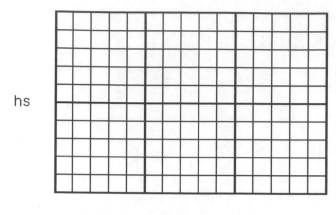

hs

WT

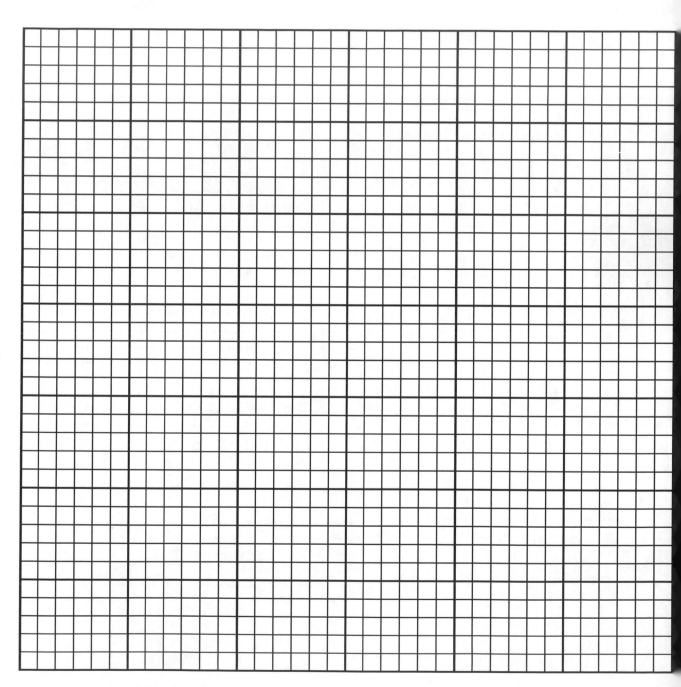

hs

WT

MERIDIAN TRANSIT SIGHT REDUCTION FORM

TIME OF TRANSIT

DR GHA _____ ° _____ . ____ '

  = Lo W, 360° - Lo E

GHA ____ h _____ ° ____ . ____ '

GHA Diff _____ ° ____ . ____ '

Min-Sec _____

UT _____

ZD $^{E-}_{W+}$ ( ) _____   (rev)

ZT _____

- -

Last sight WT _____

Difference _____

ZT + Difference _____

LAN

S _____ . ___ kt   Cn _____ °

Sn ( ) _____ . ___ kt

Sn - d ( ) _____ . ___ '/hr

  d positive if change is Northerly

ΔWT _____ m

Δ hs ( ) _____ . ___ '

WT _____

WE $^{f-}_{s+}$ ( ) _____

ZT _____

ZD $^{E-}_{W+}$ ( ) _____

UT _____

LONGITUDE

GHA ____ h _____ ° ____ . ____ '

____ m ___ s _____ ° ____ . ____ '

_____ . ____ '

360° | 359° 59 .10 '

Lo _____ ° ____ . ____ ' E

BASIC DATA

Date_____ Body _____

DR L _____ ° ____ . ____ ' $^{N}_{S}$

DR Lo _____ ° ____ . ____ $^{E}_{W}$

DECLINATION

Dec____ hr _____ ° ____ . ___ ' $^{N}_{S}$

d ( ) ____ . ____ '/hr

  d corr ( ) _____ . ____ '

Dec _____ ° ____ . ___ ' $^{N}_{S}$

LATITUDE

50°N —
|
|
EQ —
|
|
50°S —

90° | 89 ° 59 . 10 '

Co-Alt = 90° - Ho _____ ° ____ . ____ ' $^{N}_{S}$

*Co-Alt name is observer
position relative to body.*   Dec _____ ° ____ . ____ ' $^{N}_{S}$

L = Co-Alt ± Dec _____ ° ____ . ____ ' $^{N}_{S}$

*If Co-Alt and Dec opposite
names, subtract.*

ALTITUDE

Ht of eye _____ . ___ ft

hs _____ ° ____ . ____ '

IC ( ) _____ . ____ '

Dip (−) _____ . ____ '

Total ( ) _____ . ____ '

ha _____ ° ____ . ____ '

Corr ( ) _____ . ____ '

Ho _____ ° ____ . ____ '

Sn DETERMINATION

Sn		180	170	160	150	140	130	120	110	100	90
NEGATIVE	{	180	190	200	210	220	230	240	250	260	270
		360	350	340	330	320	310	300	290	280	270

Cn = TRUE COURSE

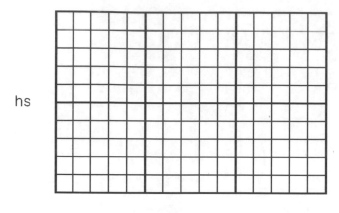

hs

WT

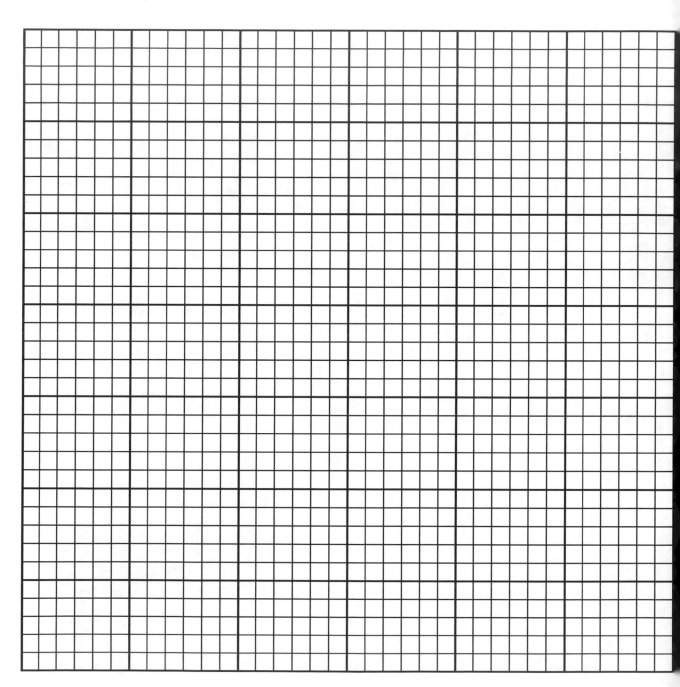

hs

WT

MERIDIAN TRANSIT SIGHT REDUCTION FORM

TIME OF TRANSIT

DR GHA _____°_____.___'

= Lo W, 360° - Lo E

GHA _____ h _____°_____.___

GHA Diff _____°_____.___

Min-Sec _____

UT _____

ZD $^{E-}_{W+}$ ( ) _____ (rev)

ZT _____

- - - - - - - - - - - - - - - - -

Last sight WT _____

Difference _____

ZT + Difference _____

LAN

S _____.___ kt   Cn _____°

Sn ( ) _____.___ kt

Sn - d ( ) _____.___'/hr

d positive if change is Northerly

ΔWT _____ m

Δ hs ( ) _____.___'

WT _____

WE $^{f-}_{s+}$ ( ) _____

ZT _____

ZD $^{E-}_{W+}$ ( ) _____

UT _____

LONGITUDE

GHA _____ h _____°_____.___'

_____ m _____ s _____°_____.___'

Lo _____°_____.___'

360°   359° 59 .10 '

Lo _____°_____.___' E

BASIC DATA

Date _____ Body _____

DR L _____°_____.___' ^N_S

DR Lo _____°_____.___' $^{E}_{W}$

DECLINATION

Dec _____ hr _____°_____.___' ^N_S

d ( ) _____.___'/hr

d corr ( ) _____.___'

Dec _____°_____.___' ^N_S

LATITUDE

50°N

EQ

50°S

LATITUDE   90°   89 ° 59 . 10 '

Co-Alt = 90° - Ho _____°_____.___' ^N_S

Co-Alt name is observer position relative to body.   Dec _____°_____.___' ^N_S

L = Co-Alt ± Dec _____°_____.___' ^N_S

If Co-Alt and Dec opposite names, subtract.

ALTITUDE

Ht of eye _____.___ ft

hs _____°_____.___

IC ( ) _____.___'

Dip (−) _____.___'

Total ( ) _____.___'

ha _____°_____.___

Corr ( ) _____.___'

Ho _____°_____.___

Sn DETERMINATION

Sn	180	170	160	150	140	130	120	110	100	90
NEGATIVE	180	190	200	210	220	230	240	250	260	270
	360	350	340	330	320	310	300	290	280	270

Cn = TRUE COURSE

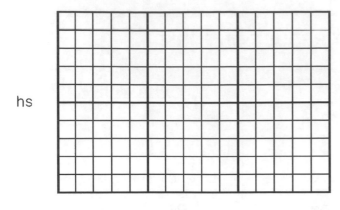

hs

WT

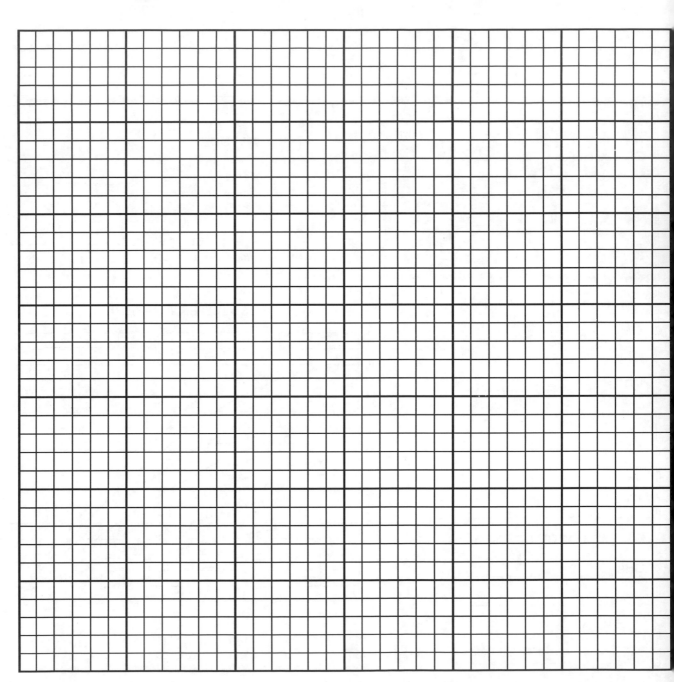

hs

WT

MERIDIAN TRANSIT SIGHT REDUCTION FORM

TIME OF TRANSIT

DR GHA _____°_____.____'

= Lo W, 360° - Lo E

GHA _____h _____°_____.____'

GHA Diff _____°_____.____'

Min-Sec _____

UT _____

ZD $^{E-}_{W+}$ ( )_____ (rev)

ZT _____

- - - - - - - - - - - - - - - - - - - -

Last sight WT _____

Difference _____

ZT + Difference _____

LAN

S _____.___ kt   Cn _____°

Sn ( ) _____.___ kt

Sn - d ( ) _____.___ '/hr

d positive if change is Northerly

ΔWT _____ m

Δ hs ( )_____.____'

WT _____

WE $^{f-}_{s+}$ ( ) _____

ZT _____

ZD $^{E-}_{W+}$ ( )_____

UT _____

LONGITUDE

GHA _____h _____°_____.____'

_____m ___s _____°_____.____'

360° _____359°__59_.10_'

Lo _____°_____.____' E

BASIC DATA

Date_____ Body _____

DR L _____°_____.____ ^N_S

DR Lo _____°_____.____ $^{S}_{E}_{W}$

DECLINATION

Dec_____hr _____°_____.____' ^N_S

d ( )_____.____'/hr

d corr ( ) _____.____'

Dec _____°_____.____' ^N_S

50°N

EQ

50°S

LATITUDE

90° ____89_°_59_.10_'

Co-Alt = 90° - Ho _____°_____.____ ^N_S

Co-Alt name is observer position relative to body.

Dec _____°_____.____' ^N_S

L = Co-Alt ± Dec _____°_____.____' ^N_S

If Co-Alt and Dec opposite names, subtract.

ALTITUDE

Ht of eye _____.____ ft

hs _____°_____.____'

IC ( ) _____.____

Dip (−) _____.____

Total ( ) _____.____

ha _____°_____.____'

Corr ( ) _____.____

Ho _____°_____.____'

Sn DETERMINATION

Sn, knots

Sn	180	170	160	150	140	130	120	110	100	90
NEGATIVE {	180	190	200	210	220	230	240	250	260	270
	360	350	340	330	320	310	300	290	280	270

Cn = TRUE COURSE

hs

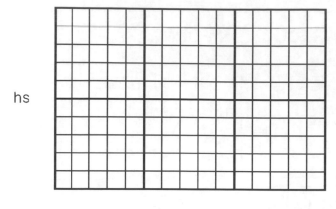

WT

hs

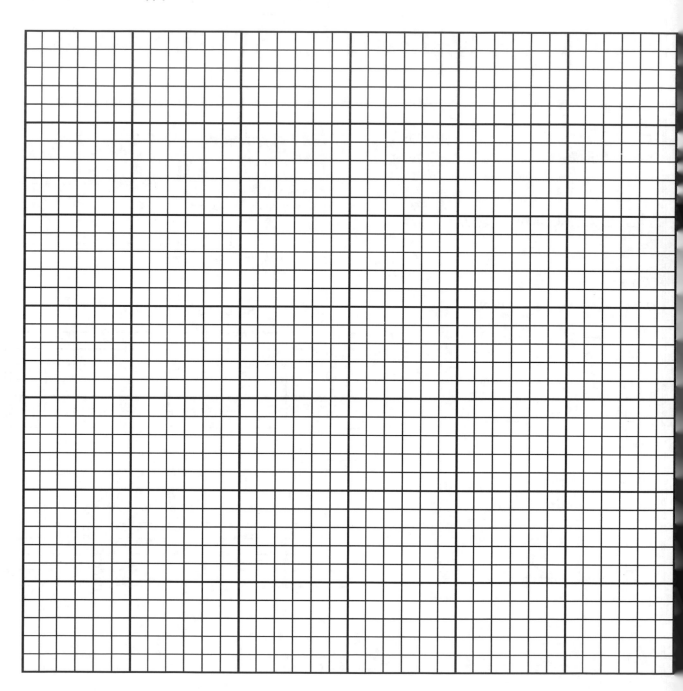

WT

MERIDIAN TRANSIT SIGHT REDUCTION FORM

TIME OF TRANSIT

DR GHA _____°____.____'

  = Lo W, 360° - Lo E

GHA____h_____°____.____'

GHA Diff _____°____.____'

Min-Sec _____

UT _____

ZD $\begin{matrix} E- \\ W+ \end{matrix}$ ( )_____ (rev)

ZT _____

- -

Last sight WT _____

Difference _____

ZT + Difference _____

LAN

S ____.___ kt   Cn _____°

Sn ( ) ____.___ kt

Sn - d ( ) _____.___'/hr

  d positive if change is Northerly

ΔWT _____ m

Δ hs ( )_____.___'

WT _____

WE $\begin{matrix} f- \\ s+ \end{matrix}$ ( ) _____

ZT _____

ZD $\begin{matrix} E- \\ W+ \end{matrix}$ ( )_____

UT _____

LONGITUDE

GHA____h____°____.____'

____m____s____°____.____'

360° ___279° 59 .10 '

Lo ____°____.____' E

BASIC DATA

Date_____ Body_____

DR L _____°____.____' $\begin{matrix} N \\ S \end{matrix}$

DR Lo _____°____.____' $\begin{matrix} E \\ W \end{matrix}$

DECLINATION

Dec____hr_____°____.____' $\begin{matrix} N \\ S \end{matrix}$

d ( )____.____'/hr

  d corr ( ) ____.____'

Dec _____°____.____' $\begin{matrix} N \\ S \end{matrix}$

ALTITUDE

Ht of eye _____.___ ft

hs _____°____.____'

IC ( ) ____.____'

Dip (−) ____.____'

Total ( ) ____.____'

ha _____°____.____'

Corr ( ) ____.____'

Ho _____°____.____'

LATITUDE

90°   89 ° 59 .10 '

Co-Alt = 90° - Ho _____°____.____' $\begin{matrix} N \\ S \end{matrix}$

Co-Alt name is observer position relative to body.

Dec _____°____.____' $\begin{matrix} N \\ S \end{matrix}$

L = Co-Alt ± Dec _____°____.____' $\begin{matrix} N \\ S \end{matrix}$

If Co-Alt and Dec opposite names, subtract.

Sn DETERMINATION

Sn {	180	170	160	150	140	130	120	110	100	90
NEGATIVE {	180	190	200	210	220	230	240	250	260	270
	360	350	340	330	320	310	300	290	280	270

Cn = TRUE COURSE

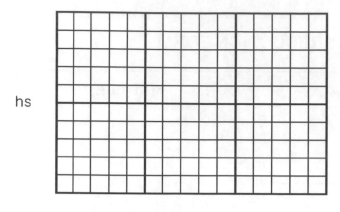

hs

WT

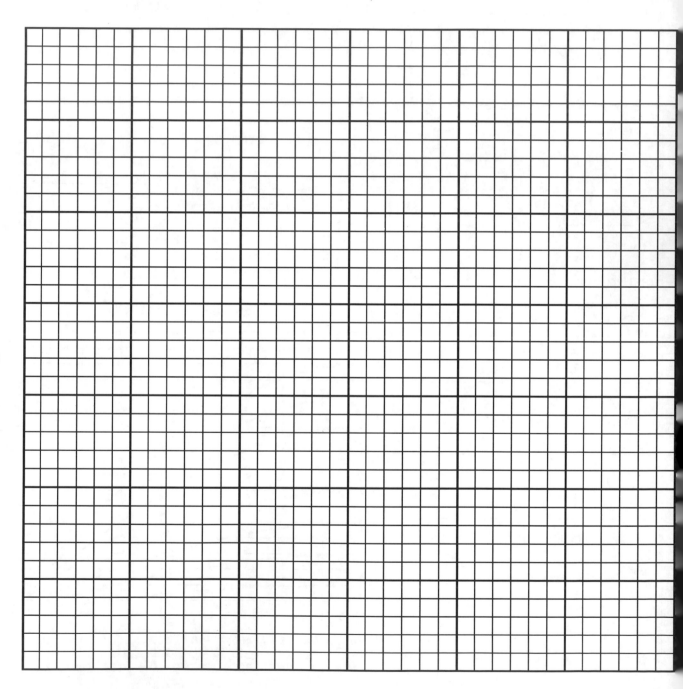

hs

WT

MERIDIAN TRANSIT SIGHT REDUCTION FORM

TIME OF TRANSIT

DR GHA _____°_____.___'

= Lo W, 360° - Lo E

GHA_____h_____°_____.___'

GHA Diff _____°_____.___'

Min-Sec _____

UT _____

ZD $^{E-}_{W+}$ ( )_____ (rev)

ZT _____

- - - - - - - - - - - - - - - - -

Last sight WT _____

Difference _____

ZT + Difference _____

LAN

S _____.___ kt   Cn _____°

Sn ( ) _____.___ kt

Sn - d ( ) _____.___'/hr

d positive if change is Northerly

ΔWT _____ m

Δ hs ( ) _____.___'

WT _____

WE $^{f-}_{s+}$ ( ) _____

ZT _____

ZD $^{E-}_{W+}$ ( )_____

UT _____

LONGITUDE

GHA _____h _____°_____.___'

_____m _____s _____°_____.___'

360° _____ 359° 59 .10'

Lo _____°_____.___' E

BASIC DATA

Date_____ Body _____

DR L _____°_____.___' ^N_S

DR Lo _____°_____.___' ^E_W

DECLINATION

Dec_____hr _____°_____.___' ^N_S

d ( )_____.___'/hr

d corr ( ) _____.___'

Dec _____°_____.___' ^N_S

Sn DETERMINATION

Sn, knots

Sn {180	170	160	150	140	130	120	110	100	90
NEGATIVE {180	190	200	210	220	230	240	250	260	270
360	350	340	330	320	310	300	290	280	270

Cn = TRUE COURSE

ALTITUDE

Ht of eye _____.___ ft

hs _____°_____.___'

IC ( ) _____.___

Dip (−) _____.___'

Total ( ) _____.___'

ha _____°_____.___'

Corr ( ) _____.___'

Ho _____°_____.___'

LATITUDE

90° _____ 89° 59 .10'

Co-Alt = 90° - Ho _____°_____.___' ^N_S

Co-Alt name is observer position relative to body.

Dec _____°_____.___' ^N_S

L = Co-Alt ± Dec _____°_____.___' ^N_S

If Co-Alt and Dec opposite names, subtract.

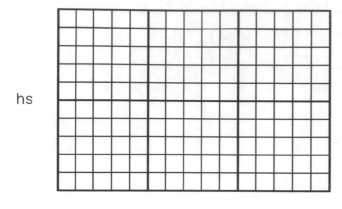

hs

WT

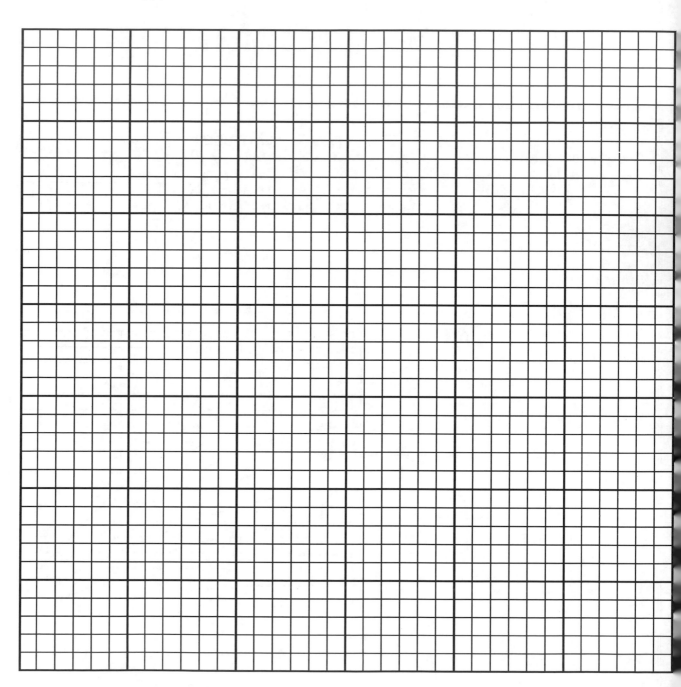

hs

WT

MERIDIAN TRANSIT SIGHT REDUCTION FORM

TIME OF TRANSIT

DR GHA _____°_____.___'

  = *Lo W, 360° - Lo E*

GHA ____ h _____°_____.___'

GHA Diff _____°_____.___'

Min-Sec _____

UT _____

ZD $^{E-}_{W+}$ ( ) _____ (rev)

ZT _____

- -

Last sight WT _____

Difference _____

ZT + Difference _____

LAN

S _____.___ kt   Cn _____°

Sn ( ) _____.___ kt

Sn - *d* ( ) _____.___'/hr

  d positive if change is Northerly

ΔWT _____ m

Δ hs ( ) _____.___'

WT _____

WE $^{f-}_{s+}$ ( ) _____

ZT _____

ZD $^{E-}_{W+}$ ( ) _____

UT _____

LONGITUDE

GHA ____ h _____°_____.___'

____ m ____ s _____°_____.___'

360° _____359°_59_.10'

Lo _____°_____.___' E

BASIC DATA

Date_____ Body_____

DR L _____°_____.___' $^{N}_{S}$

DR Lo _____°_____.___' $^{E}_{W}$

DECLINATION

Dec ____ hr _____°_____.___' $^{N}_{S}$

d ( ) _____.___'/hr

  d corr ( ) _____.___'

Dec _____°_____.___' $^{N}_{S}$

LATITUDE

90° _____89_°_59_.10'

Co-Alt = 90° - Ho _____°_____.___' $^{N}_{S}$

Co-Alt name is observer
position relative to body.   Dec _____°_____.___' $^{N}_{S}$

L = Co-Alt ± Dec _____°_____.___' $^{N}_{S}$

If Co-Alt and Dec opposite
names, subtract.

(scale: 50°N — EQ — 50°S)

ALTITUDE

Ht of eye _____.___ ft

hs _____°_____.___'

IC ( ) _____.___'

Dip (−) _____.___'

Total ( ) _____.___'

ha _____°_____.___'

Corr ( ) _____.___'

Ho _____°_____.___'

Sn DETERMINATION

Sn $\{$	180	170	160	150	140	130	120	110	100	90
NEGATIVE $\{$	180	190	200	210	220	230	240	250	260	270
	360	350	340	330	320	310	300	290	280	270

Cn = TRUE COURSE

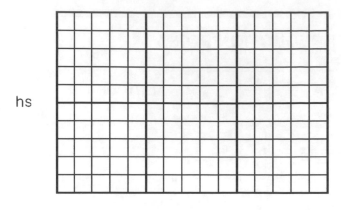

hs

WT

hs

WT

MERIDIAN TRANSIT SIGHT REDUCTION FORM

TIME OF TRANSIT

DR GHA _____°_____.___'

  = *Lo W, 360° - Lo E*

GHA____ h_____°_____.___'

GHA Diff _____°_____.___'

Min-Sec _____

UT _____

ZD $^{E-}_{W+}$ ( ) _____ (rev)

ZT _____

- -

Last sight WT _____

Difference _____

ZT + Difference _____

LAN

S _____.___ kt   Cn _____°

Sn ( ) _____.___ kt

Sn - *d* ( ) _____.___'/hr

  d positive if change is Northerly

ΔWT _____ m

Δ hs ( ) _____.___'

WT   _____

WE $^{f-}_{s+}$ ( ) _____

ZT _____

ZD $^{E-}_{W+}$ ( ) _____

UT _____

LONGITUDE

GHA ____ h _____°_____.___'

____ m ___ s _____°_____.___'

360°   359° 59 .10 '

Lo _____°_____.___' E

BASIC DATA

Date_____ Body _____

DR L _____°_____.___' $^{N}_{S}$

DR Lo _____°_____.___' $^{E}_{W}$

DECLINATION

Dec____ hr_____°_____.___' $^{N}_{S}$

d ( ) _____.____'/hr

  d corr ( ) _____.___'

Dec _____°_____.___' $^{N}_{S}$

LATITUDE

           90° _____ 89 ° 59 .10 '

Co-Alt = 90° - Ho _____°_____.___' $^{N}_{S}$

Co-Alt name is observer position relative to body.   Dec _____°_____.___' $^{N}_{S}$

L = Co-Alt ± Dec _____°_____.___' $^{N}_{S}$

If Co-Alt and Dec opposite names, subtract.

ALTITUDE

Ht of eye _____.___ ft

hs _____°_____.___'

IC ( ) _____.___'

Dip (−) _____.___'

Total ( ) _____.___'

ha _____°_____.___'

Corr ( ) _____.___'

Ho _____°_____.___'

Sn DETERMINATION

Sn	180	170	160	150	140	130	120	110	100	90
NEGATIVE	180	190	200	210	220	230	240	250	260	270
	360	350	340	330	320	310	300	290	280	270

Cn = TRUE COURSE

hs

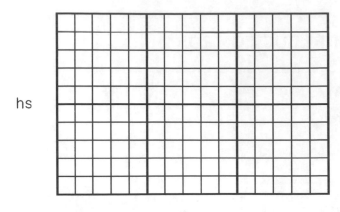

WT

hs

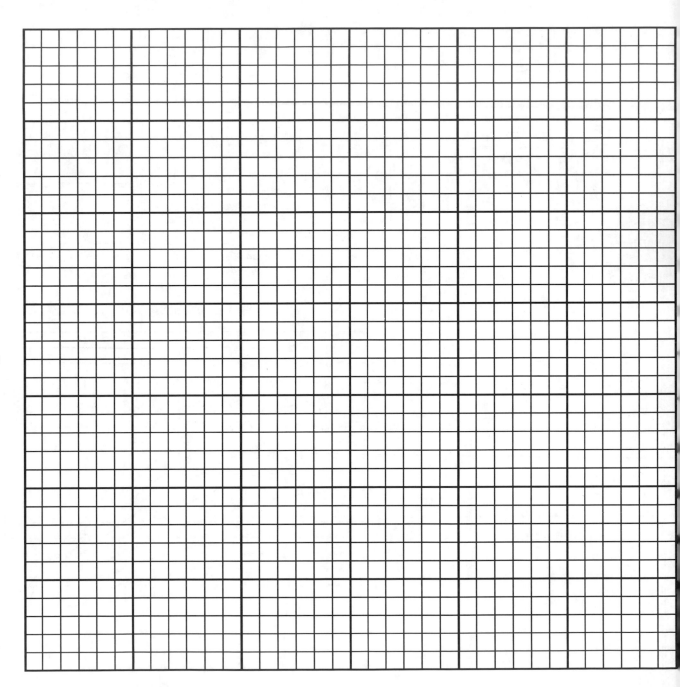

WT

MERIDIAN TRANSIT SIGHT REDUCTION FORM

TIME OF TRANSIT

DR GHA _____ ° ____ . ___ '

= Lo W, 360° - Lo E

GHA ____ h _____ ° ____ . ___

GHA Diff _____ ° ____ . ___

Min-Sec _____

UT _____

ZD $\begin{smallmatrix}E-\\W+\end{smallmatrix}$ ( ) _____ (rev)

ZT _____

- - - - - - - - - - - - - - - -

Last sight WT _____

Difference _____

ZT + Difference _____

LAN

S _____ . ___ kt   Cn _____ °

Sn ( ) _____ . ___ kt

Sn - d ( ) _____ . ___ '/hr

d positive if change is Northerly

ΔWT _____ m

Δ hs ( ) _____ . ___ '

WT _____

WE $\begin{smallmatrix}f-\\s+\end{smallmatrix}$ ( ) _____

ZT _____

ZD $\begin{smallmatrix}E-\\W+\end{smallmatrix}$ ( ) _____

UT _____

LONGITUDE

GHA ____ h _____ ° ____ . ___

____ m ____ s _____ ° ____ . ___

360° _____ 359° 59 . 10 '

Lo _____ ° ____ . ___ ' E

BASIC DATA

Date_____   Body _____

DR L _____ ° ____ . ___ ' $\begin{smallmatrix}N\\S\end{smallmatrix}$

DR Lo _____ ° ____ . ___ $\begin{smallmatrix}E\\W\end{smallmatrix}$

DECLINATION

Dec ____ hr _____ ° ____ . ___ ' $\begin{smallmatrix}N\\S\end{smallmatrix}$

d ( ) ____ . ___ '/hr

d corr ( ) _____ . ___ '

Dec _____ ° ____ . ___ ' $\begin{smallmatrix}N\\S\end{smallmatrix}$

ALTITUDE

Ht of eye _____ . ___ ft

hs _____ ° ____ . ___

IC ( ) _____ . ___

Dip (−) _____ . ___

Total ( ) _____ . ___

ha _____ ° ____ . ___

Corr ( ) _____ . ___ '

Ho _____ ° ____ . ___ '

LATITUDE

90°   89 ° 59 . 10 '

Co-Alt = 90° - Ho _____ ° ____ . ___ $\begin{smallmatrix}N\\S\end{smallmatrix}$

Co-Alt name is observer position relative to body.   Dec _____ ° ____ . ___ ' $\begin{smallmatrix}N\\S\end{smallmatrix}$

L = Co-Alt ± Dec _____ ° ____ . ___ ' $\begin{smallmatrix}N\\S\end{smallmatrix}$

If Co-Alt and Dec opposite names, subtract.

Sn DETERMINATION

Sn	180	170	160	150	140	130	120	110	100	90
NEGATIVE	180	190	200	210	220	230	240	250	260	270
	360	350	340	330	320	310	300	290	280	270

Cn = TRUE COURSE

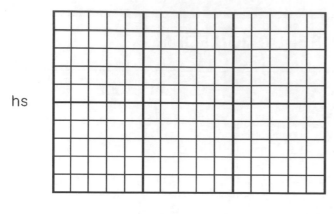

WT

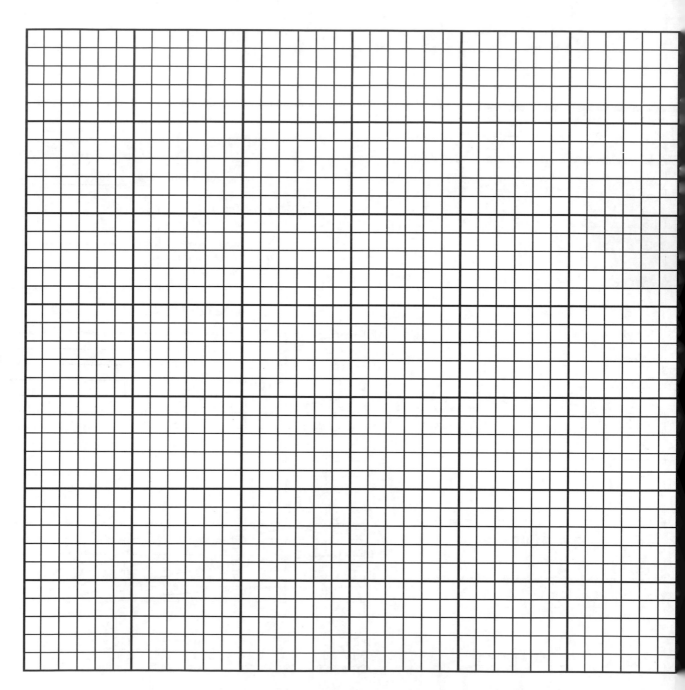

WT

MERIDIAN TRANSIT SIGHT REDUCTION FORM

TIME OF TRANSIT

DR GHA _____ ° _____ . ___ '

= Lo W, 360° - Lo E

GHA _____ h _____ ° ___ . ___ '

GHA Diff _____ ° ___ . ___ '

Min-Sec _____

UT _____

ZD $\begin{smallmatrix} E- \\ W+ \end{smallmatrix}$ ( ) _____ (rev)

ZT _____

- - - - - - - - - - - - - - - - - - - -

Last sight WT _____

Difference _____

ZT + Difference _____

LAN

S _____ . ___ kt   Cn _____ °

Sn ( ) _____ . ___ kt

Sn - d ( ) _____ . ___ '/hr

d positive if change is Northerly

ΔWT _____ m

Δ hs ( ) _____ . ___ '

WT _____

WE $\begin{smallmatrix} f- \\ s+ \end{smallmatrix}$ ( ) _____

ZT _____

ZD $\begin{smallmatrix} E- \\ W+ \end{smallmatrix}$ ( ) _____

UT _____

LONGITUDE

GHA _____ h _____ ° ___ . ___ '

___ m ___ s _____ ° ___ . ___ '

_____

360° 359° 59 . 10 '

Lo _____ ° ___ . ___ ' E

BASIC DATA

Date _____   Body _____

DR L _____ ° ___ . ___ ' $\begin{smallmatrix} N \\ S \end{smallmatrix}$

DR Lo _____ ° ___ . ___ ' $\begin{smallmatrix} E \\ W \end{smallmatrix}$

DECLINATION

Dec _____ hr _____ ° ___ . ___ ' $\begin{smallmatrix} N \\ S \end{smallmatrix}$

d ( ) ___ . ___ '/hr

d corr ( ) _____ . ___ '

Dec _____ ° ___ . ___ ' $\begin{smallmatrix} N \\ S \end{smallmatrix}$

LATITUDE

50°N ⊤

EQ ⊣

50°S ⊥

90°   89 ° 59 . 10 '

Co-Alt = 90° - Ho _____ ° ___ . ___ ' $\begin{smallmatrix} N \\ S \end{smallmatrix}$

Co-Alt name is observer position relative to body.   Dec _____ ° ___ . ___ ' $\begin{smallmatrix} N \\ S \end{smallmatrix}$

L = Co-Alt ± Dec _____ ° ___ . ___ ' $\begin{smallmatrix} N \\ S \end{smallmatrix}$

If Co-Alt and Dec opposite names, subtract.

ALTITUDE

Ht of eye _____ . ___ ft

hs _____ ° ___ . ___ '

IC ( ) _____ . ___ '

Dip (−) _____ . ___ '

Total ( ) _____ . ___ '

ha _____ ° ___ . ___ '

Corr ( ) _____ . ___ '

Ho _____ ° ___ . ___ '

Sn DETERMINATION

Sn NEGATIVE {	180	170	160	150	140	130	120	110	100	90
	180	190	200	210	220	230	240	250	260	270
	360	350	340	330	320	310	300	290	280	270

Cn = TRUE COURSE

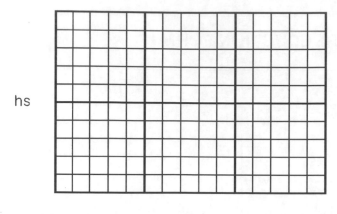

hs

WT

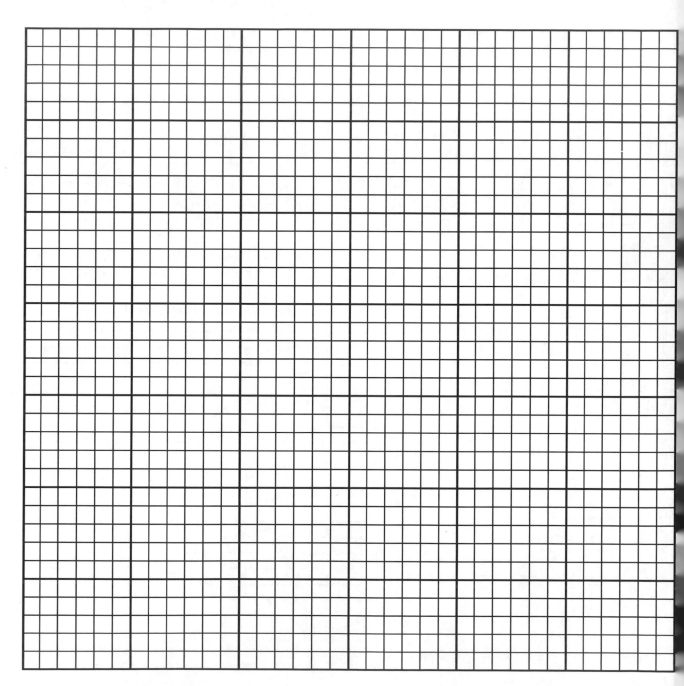

hs

WT

MERIDIAN TRANSIT SIGHT REDUCTION FORM

TIME OF TRANSIT

DR GHA _____ ° ____ . ___ '

  = Lo W, 360° - Lo E

GHA ____ h _____ ° ____ . ___ '

GHA Diff _____ ° ____ . ___ '

Min-Sec _____

UT _____

ZD $^{E-}_{W+}$ ( ) _____ (rev)

ZT _____

- -

Last sight WT _____

Difference _____

ZT + Difference _____

LAN

S ____ . __ kt   Cn _____ °

Sn ( ) _____ . __ kt

Sn - d ( ) _____ . ___ '/hr

  d positive if change is Northerly

ΔWT _____ m

Δ hs ( ) _____ . ___ '

WT _____

WE $^{f-}_{s+}$ ( ) _____

ZT _____

ZD $^{E-}_{W+}$ ( ) _____

UT _____

LONGITUDE

GHA ____ h _____ ° ____ . ___ '

____ m ___ s _____ ° ____ . ___ '

360°   359° 59 .10 '

Lo _____ ° ____ . ___ ' E

BASIC DATA

Date _____   Body _____

DR L _____ ° ____ . ___ ' $^{N}_{S}$

DR Lo _____ ° ____ . ___ ' $^{E}_{W}$

DECLINATION

Dec ____ hr _____ ° ____ . ___ ' $^{N}_{S}$

d ( ) ____ . ___ '/hr

  d corr ( ) _____ . ___ '

Dec _____ ° ____ . ___ ' $^{N}_{S}$

ALTITUDE

Ht of eye _____ . ___ ft

hs _____ ° ____ . ___ '

  IC ( ) _____ . ___ '

  Dip (−) _____ . ___ '

  Total ( ) _____ . ___ '

ha _____ ° ____ . ___ '

  Corr ( ) _____ . ___ '

Ho _____ ° ____ . ___ '

LATITUDE

90°   89 ° 59 . 10 '

Co-Alt = 90° - Ho _____ ° ____ . ___ ' $^{N}_{S}$

Co-Alt name is observer position relative to body.

Dec _____ ° ____ . ___ ' $^{N}_{S}$

L = Co-Alt ± Dec _____ ° ____ . ___ ' $^{N}_{S}$

If Co-Alt and Dec opposite names, subtract.

(latitude scale: 50°N — EQ — 50°S)

Sn DETERMINATION

Sn, knots (y-axis: 0, 2, 4, 6, 8, 10, 12, 14, 16, 18)

Sn $\{$	180	170	160	150	140	130	120	110	100	90
NEGATIVE $\{$	180	190	200	210	220	230	240	250	260	270
	360	350	340	330	320	310	300	290	280	270

Cn = TRUE COURSE

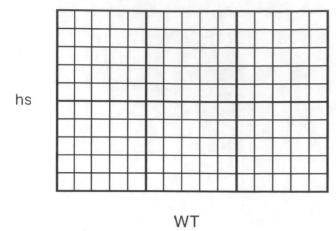

hs

WT

hs

WT

MERIDIAN TRANSIT SIGHT REDUCTION FORM

TIME OF TRANSIT

DR GHA _____°_____.___'

= Lo W, 360° - Lo E

GHA_____ h_____°_____.___'

GHA Diff _____°_____.___'

Min-Sec _____

UT _____

ZD $\begin{array}{c}E-\\W+\end{array}$ ( )_____ (rev)

ZT _____

- -

Last sight WT _____

Difference _____

ZT + Difference _____

LAN

S _____.___ kt   Cn _____°

Sn ( ) _____.___ kt

Sn - d ( ) _____.___'/hr

d positive if change is Northerly

ΔWT _____ m

Δ hs ( )_____.___'

WT _____

WE $\begin{array}{c}f-\\s+\end{array}$ ( ) _____

ZT _____

ZD $\begin{array}{c}E-\\W+\end{array}$ ( )_____

UT _____

LONGITUDE

GHA _____ h _____°_____.___'

_____ m _____ s _____°_____.___'

_____

360° _____ 359°_59_.10'

Lo _____°_____.___' E

BASIC DATA

Date_____ Body _____

DR L _____°_____.___' $\begin{array}{c}N\\S\end{array}$

DR Lo _____°_____.___' $\begin{array}{c}E\\W\end{array}$

DECLINATION

Dec_____ hr_____°_____.___' $\begin{array}{c}N\\S\end{array}$

d ( )_____.___'/hr

d corr ( ) _____.___'

Dec _____°_____.___' $\begin{array}{c}N\\S\end{array}$

LATITUDE

LATITUDE   90° _____ 89°_59_.10'

Co-Alt = 90° - Ho _____°_____.___' $\begin{array}{c}N\\S\end{array}$

Co-Alt name is observer position relative to body.   Dec _____°_____.___' $\begin{array}{c}N\\S\end{array}$

L = Co-Alt ± Dec _____°_____.___' $\begin{array}{c}N\\S\end{array}$

If Co-Alt and Dec opposite names, subtract.

ALTITUDE

Ht of eye _____.___ ft

hs _____°_____.___'

IC ( ) _____.___'

Dip (−) _____.___'

Total ( ) _____.___'

ha _____°_____.___'

Corr ( ) _____.___'

Ho _____°_____.___'

Sn DETERMINATION

Sn	180	170	160	150	140	130	120	110	100	90
NEGATIVE	180	190	200	210	220	230	240	250	260	270
	360	350	340	330	320	310	300	290	280	270

Cn = TRUE COURSE

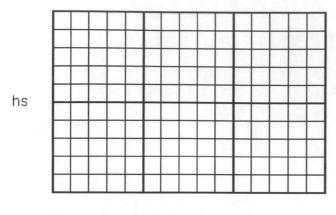

hs

WT

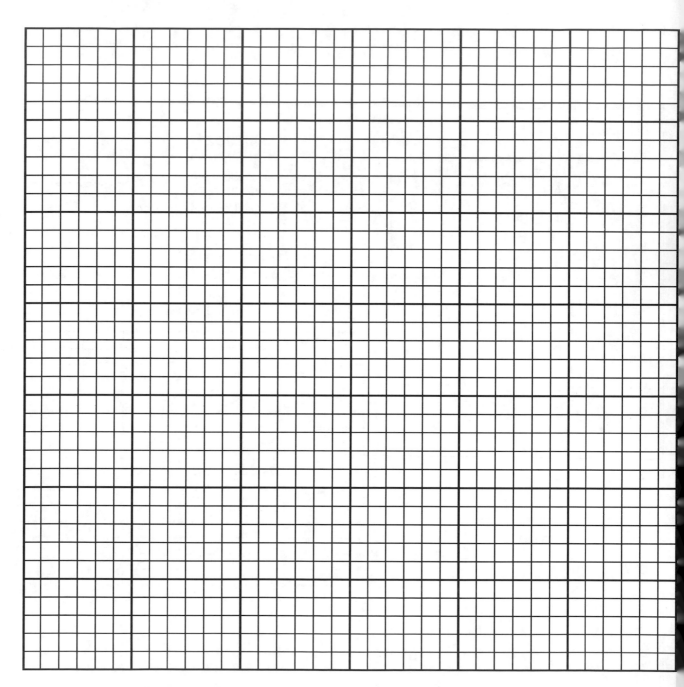

hs

WT

MERIDIAN TRANSIT SIGHT REDUCTION FORM

TIME OF TRANSIT

DR GHA _____°____.___'

= *Lo W, 360° - Lo E*

GHA ____ h _____°___.___'

GHA Diff _____°___.___'

Min-Sec _____

UT _____

ZD $^{E-}_{W+}$ ( ) _____   (rev)

ZT _____

- - - - - - - - - - - - - - - - - -

Last sight WT _____

Difference _____

ZT + Difference _____

LAN

S _____.__ kt   Cn _____°

Sn ( ) _____.__ kt

Sn - *d* ( ) _____.___'/hr

d positive if change is Northerly

ΔWT _____ m

Δ hs ( ) _____.___'

WT _____

WE $^{f-}_{s+}$ ( ) _____

ZT _____

ZD $^{E-}_{W+}$ ( ) _____

UT _____

LONGITUDE

GHA ____ h _____°____.___'

___ m ___ s _____°____.___'

360°   359° 59 .10 '

Lo _____°____.___' E

BASIC DATA

Date_____ Body _____

DR L _____°___.___' $^{N}_{S}$

DR Lo _____°___.___' $^{E}_{W}$

DECLINATION

Dec____ hr _____°___.___' $^{N}_{S}$

d ( ) ____.___'/hr

d corr ( ) _____.___'

Dec _____°___.___' $^{N}_{S}$

LATITUDE

90°   89 ° 59 . 10 '

Co-Alt = 90° - Ho _____°___.___' $^{N}_{S}$

Co-Alt name is observer position relative to body.   Dec _____°___.___' $^{N}_{S}$

L = Co-Alt ± Dec _____°___.___' $^{N}_{S}$

If Co-Alt and Dec opposite names, subtract.

ALTITUDE

Ht of eye _____.___ ft

hs _____°___.___'

IC ( ) _____.___'

Dip (−) _____.___'

Total ( ) _____.___'

ha _____°___.___'

Corr ( ) _____.___'

Ho _____°___.___'

Sn DETERMINATION

Sn	180	170	160	150	140	130	120	110	100	90
NEGATIVE	180	190	200	210	220	230	240	250	260	270
	360	350	340	330	320	310	300	290	280	270

Cn = TRUE COURSE

hs

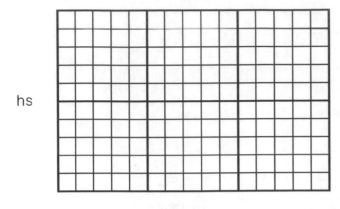

WT

hs

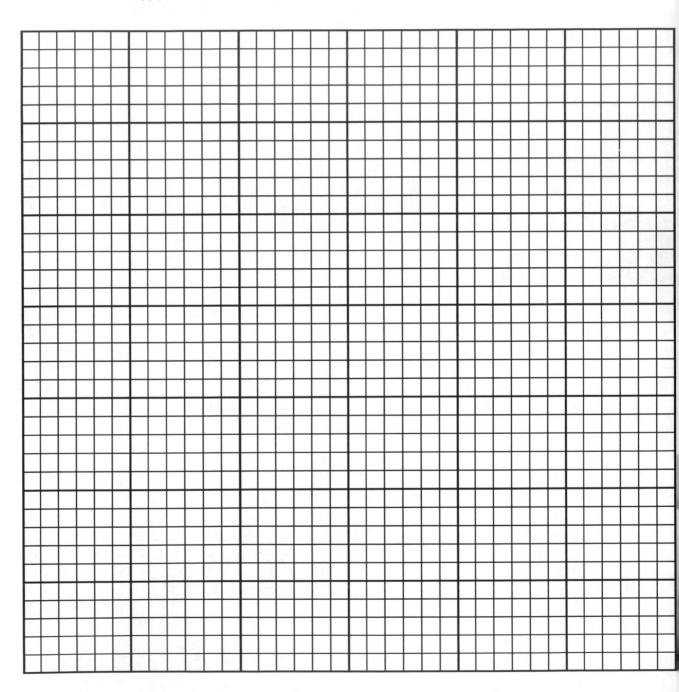

WT

MERIDIAN TRANSIT SIGHT REDUCTION FORM

TIME OF TRANSIT

DR GHA _____°_____.___'

= Lo W, 360° - Lo E

GHA_____ h_____°_____.___'

GHA Diff _____°_____.___'

Min-Sec _____

UT _____

ZD $^{E-}_{W+}$ ( ) _____ (rev)

ZT _____

- - - - - - - - - - - - - - - - - - - -

Last sight WT _____

Difference _____

ZT + Difference _____

LAN

S _____.___ kt   Cn _____°

Sn ( ) _____.___ kt

Sn - d ( ) _____.___'/hr

d positive if change is Northerly

ΔWT _____ m

Δ hs ( ) _____.___'

WT _____

WE $^{f-}_{s+}$ ( ) _____

ZT _____

ZD $^{E-}_{W+}$ ( ) _____

UT _____

LONGITUDE

GHA _____ h _____°_____.___'

_____ m _____ s _____°_____.___'

_____._____

360° 359°59.10'

Lo _____°_____.___' E

BASIC DATA

Date_____ Body_____

DR L _____°_____.___' ^N_S

DR Lo _____°_____.___' $^{S \; E}_W$

DECLINATION

Dec_____ hr _____°_____.___' ^N_S

d ( ) _____.___'/hr

d corr ( ) _____.___'

Dec _____°_____.___' ^N_S

ALTITUDE

Ht of eye _____.___ ft

hs _____°_____.___'

IC ( ) _____.___'

Dip (−) _____.___'

Total ( ) _____.___'

ha _____°_____.___'

Corr ( ) _____.___'

Ho _____°_____.___'

LATITUDE

90°   89°59.10'

Co-Alt = 90° - Ho _____°_____.___' ^N_S

Co-Alt name is observer position relative to body.

Dec _____°_____.___' ^N_S

L = Co-Alt ± Dec _____°_____.___' ^N_S

If Co-Alt and Dec opposite names, subtract.

Sn DETERMINATION

Sn	180	170	160	150	140	130	120	110	100	90
NEGATIVE	180	190	200	210	220	230	240	250	260	270
	360	350	340	330	320	310	300	290	280	270

Cn = TRUE COURSE

hs

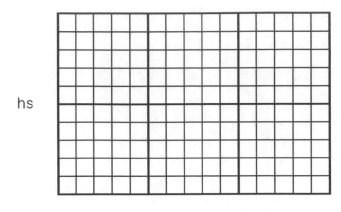

WT

hs

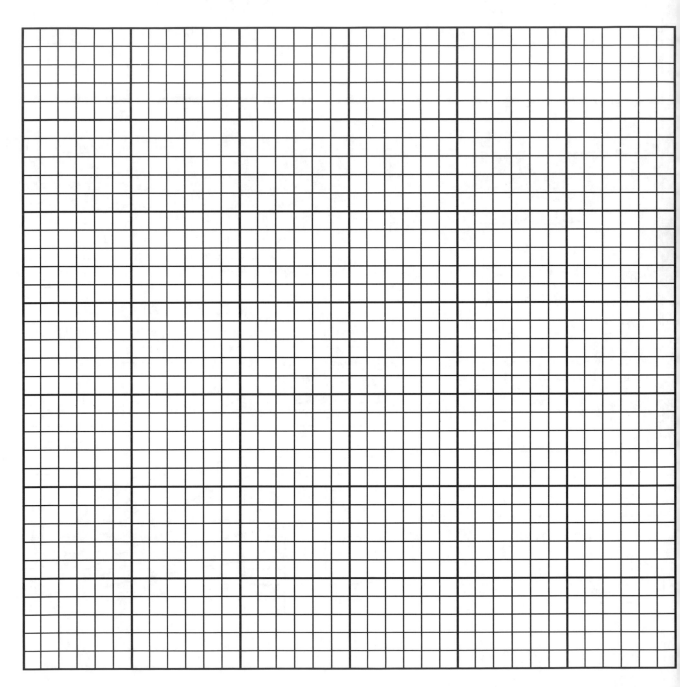

WT